MATLAB - PROFESSIONAL APPLICATIONS IN POWER SYSTEM

Edited by **Ali Saghafinia**

MATLAB - Professional Applications in Power System

http://dx.doi.org/10.5772/intechopen.68720

Edited by Ali Saghafinia

Contributors

Alexandru Baloi, Adrian Pana, G D Anbarasi Jebaselvi, V Meenakshi, Chiemela Onunka, Evans E Ojo, Ali Saghafinia

Notice

Statements and opinions expressed in the chapters are these of the individual contributors and not necessarily those of the editors or publisher. No responsibility is accepted for the accuracy of information contained in the published chapters. The publisher assumes no responsibility for any damage or injury to persons or property arising out of the use of any materials, instructions, methods or ideas contained in the book.

First published in London, United Kingdom, 2018 by IntechOpen

IntechOpen is the global imprint of INTECHOPEN LIMITED, registered in England and Wales, registration number: 11086078, The Shard, 25th floor, 32 London Bridge Street

London, SE19SG – United Kingdom

Printed in Croatia

British Library Cataloguing-in-Publication Data
A catalogue record for this book is available from the British Library

Additional hard copies can be obtained from orders@intechopen.com

MATLAB - Professional Applications in Power System, Edited by Ali Saghafinia

p. cm.

Print ISBN 978-1-78923-706-1

Online ISBN 978-1-78923-707-8

We are IntechOpen,
the world's leading publisher of
Open Access books
Built by scientists, for scientists

3,700+
Open access books available

115,000+
International authors and editors

119M+
Downloads

Our authors are among the

151
Countries delivered to

Top 1%
most cited scientists

12.2%
Contributors from top 500 universities

CLARIVATE ANALYTICS
BOOK
CITATION
INDEX
INDEXED

WEB OF SCIENCE™

Selection of our books indexed in the Book Citation Index
in Web of Science™ Core Collection (BKCI)

Interested in publishing with us?
Contact book.department@intechopen.com

Numbers displayed above are based on latest data collected.
For more information visit www.intechopen.com

Meet the editor

Ali Saghafinia was born in Esfahan, Iran, in 1973. He received a BSc degree in Electronic Engineering from Najafabad Branch, Islamic Azad University, Iran, in 1995, and an MSc degree in Electrical Engineering was awarded from Isfahan University of Technology (IUT) in 2001. He was a lecturer at the Department of Electrical Engineering, Majlesi Branch, Islamic Azad University, Esfahan, Iran, during 2002–2008. His PhD degree in Electrical Engineering was awarded in 2013 by the University of Malaya, Kuala Lumpur, Malaysia. Dr. Saghafinia passed a postdoctoral research fellowship at the UM Power Energy Dedicated Advanced Centre, University of Malaya, in 2013–2014. He also worked at the Department of Electrical and Computer Engineering, IUT, as a postdoctoral research fellow during 2016–2017. Currently, Dr. Saghafinia is an assistant professor at Majlesi Branch, Islamic Azad University. His research interests include smart grids, power electronics, electric motor drives, machine design, fault detection, and industrial engineering. He possesses 15 years of teaching experience and has authored or coauthored over 50 books, book chapters, and papers in international journals and conferences.

Contents

Chapter 1

Introductory Chapter: MATLAB Applications in Power System

Ali Saghafinia

Additional information is available at the end of the chapter

http://dx.doi.org/10.5772/intechopen.79777

1. Introduction

One of the most widespread simulation software is MathWorks MATLAB/Simulink package [1]. The MathWorks MATLAB allows the user to analyze complex static systems that are to be modeled through "numeric computation and visualization." Also, Simulink allows systems to be simulated dynamically and allows a controller to be modeled with the aid of block diagrams. Moreover, the user is allowed to concentrate on the model, rather than its implementation [2].

The simulation of power engineering applications conventionally can be a challenge for both undergraduate and postgraduate levels [1]. To easy implementation for several kinds of power structure and control structures of power engineering applications, some simulators such as MATLAB/(Simulink and coding) to be necessary—especially for students—to develop and test various circuits and controllers for power engineering, which include all branches in power engineering area [3]. This book includes some chapters to show how to simulate and work with MATLAB software for several MATLAB professional applications of power system engineering. Moreover, this book presents some techniques to simulate power matters in an easy way using the related toolbox existing in the MATLAB/Simulink and teach the simulation in the mentioned area.

2. Summary of second chapter

To facilitate the mentioned goals in Introduction Section , Chapter 2 written by Alexandru Băloi and Adrian Pană proposes a method for network harmonic impedance determination, practically. The network harmonic impedance has been used as a tool to predict the harmonic condition amplifications in the case of reactive power compensation by capacitor banks [4]. Based on the measured network harmonic impedance, a quick method has been developed to anticipate the harmonic voltage and current amplifications. Amplification factors have been calculated depending on the equivalent harmonic impedance of the network seen in the compensation bus [5]. A distribution network containing harmonics has been modeled, and then harmonic impedance has been determined in different operating conditions using MATLAB Simulink. Using the measured values and the capacitive reactance of the capacitor bank, the amplification of the harmonic voltages and currents has been estimated by calculus [6, 7]. To validate the method, the obtained results have been compared with the values obtained by simulation after the connection of the capacitor bank to the network. The chapter has proved that the network harmonic impedance is a useful tool to estimate the harmonic amplification caused by power factor correction using shunt capacitor banks.

3. Summary of third chapter

The mentioned goals are followed by Chapter 3, where Anbarasi Jebaselvi and Meenakshi present modeling, simulation and analysis of permanent magnet synchronous generator (PMSG)-based wind energy conversion systems for both open- and closed-loop control strategy [8, 9]. Since the choice of wind turbine generator and optimum power from it can be achieved by matching the load and power lines close to each other [10], maximum power from wind using PMSG has been made possible by using intelligent controllers, namely fuzzy logic controllers. As PMSG suffers from strong decay of magnetic field, which tends to reduce the generated voltage at their stator terminals drastically with load, hence not much suitable for isolated operation and thus the whole analysis has been done with grid connected network. The other major limitation includes loss of flexibility in field flux control, and thus intelligent technique like fuzzy logic mechanism has been attempted along with Space Vector Modulation (SVM) to have a smooth control of field flux and load power management in PMSG. MATLAB Simulink has been used to simulate and analyze the PMSG-based wind energy conversion systems for both open- and closed-loop control strategies [11]. It is confirmed that sufficient revenue could be generated from the proposed model by implementing the same in real-time.

4. Summary of forth chapter

To follow the mentioned goals in Introduction Section, Chapter 4, Chiemela, Onunka and Evans Eshiemogie, Ojo develop and implement a wind-induced high voltage transmission line vibration using finite element method (FEM) in MATLAB to investigate wind-induced vibration. The chapter has focused toward the development of a finite element method and its implementation on the MATLAB software. The developed finite element method (FEM) has been done based on the concept of the simply supported beam model and it has been used in modeling the transverse vibration of power line conductors [12]. First, the FEM model has been used to develop the equation of motion of the power line conductor. In addition, dampers, conditions for damping, and free- and forced vibrations of the overhead conductor have been considered in the FEM model [13]. Wind-induced experiments have been conducted in the laboratory using an actual overhead power conductor. The developed models have been simulated in the MATLAB computing environment. The results from the MATLAB simulation, finite element, and experimental recordings have been compared in order to determine the accuracy of the models and evaluate the efficacy of models simulated in MATLAB and developed using the FEM. Finally, the developed FEM has been used as the means to verify the effect of varying the conductor axial tension on the natural frequencies of the conductors.

Author details

Ali Saghafinia

Address all correspondence to: saghafi_ali@yahoo.com

Department of Electrical Engineering, Majlesi Branch, Islamic Azad University, Majlesi, Iran

References

[1] Saghafinia A, Ping HW, Uddin MN, Amindoust A. Teaching of simulation an adjustable speed drive of induction motor using MATLAB/Simulink in advanced electrical machine laboratory. Procedia-Social and Behavioral Sciences. 2013;**103**:912-921

[2] Saghafinia A, Amindoust A. Development of fuzzy applications for high performance induction motor drive. In: Induction Motors-Applications, Control and Fault Diagnostics. Rijeka, Croatia: InTech; 2015

[3] MathWorks. SIMULINK for Technical Computing. 2015. Avaliable from: http://www.mathworks.com

[4] Hui J, Yang H, Lin S, Ye M. Assessing utility harmonic impedance based on the covariance characteristic of random vectors. IEEE Transactions on Power Delivery. 2010;**25**:1778-1786

[5] Wu C-J, Chiang J-C, Yen S-S, Liao C-J, Yang J-S, Guo T-Y. Investigation and mitigation of harmonic amplification problems caused by single-tuned filters. IEEE Transactions on Power Delivery. 1998;**13**:800-806

[6] Xu W, Ahmed EE, Zhang X, Liu X. Measurement of network harmonic impedances: Practical implementation issues and their solutions. IEEE Transactions on Power Delivery. 2002;**17**:210-216

[7] Xu W, Ahmed E, Zhang X, Liu X. Measurement of network harmonic impedences: Practical implementation issues and their solutions. IEEE Power Engineering Review. 2001;**21**:63

[8] Shariatpanah H, Fadaeinedjad R, Rashidinejad M. A new model for PMSG-based wind turbine with yaw control. IEEE Transactions on Energy Conversion. 2013;**28**:929-937

[9] Melício R, Mendes VM, Catalão JP. Wind turbines with permanent magnet synchronous generator and full-power converters: modelling, control and simulation. In: Wind Turbines. Rijeka, Croatia: InTech; 2011

[10] Chinchilla M, Arnaltes S, Burgos JC. Control of permanent-magnet generators applied to variable-speed wind-energy systems connected to the grid. IEEE Transactions on Energy Conversion. 2006;**21**:130-135

[11] Harrabi N, Souissi M, Aitouche A, Chabaane M. Intelligent control of wind conversion system based on PMSG using TS fuzzy scheme. International Journal of Renewable Energy Research. 2015;**5**:952-960

[12] Lalonde S, Guilbault R, Langlois S. Numerical analysis of ACSR conductor–clamp systems undergoing wind-induced cyclic loads. IEEE Transactions on Power Delivery. 2018; **33**:1518-1526

[13] Xie T, Peng Z, Zhou Z. Study on optimization of anti-corona properties of 330-kV dampers. IEEE Transactions on Power Delivery. 2015;**30**:1827-1832

MatLab Simulink Modeling for Network-Harmonic Impedance Assessment: Useful Tool to Estimate Harmonics Amplification

Alexandru Băloi and Adrian Pană

Additional information is available at the end of the chapter

http://dx.doi.org/10.5772/intechopen.76461

Abstract

The importance of the subject is given by the fact that harmonics are making their presence felt in electrical distribution networks, and the cheapest and most widespread solution for power factor correction is the capacitor banks. This chapter proves that the harmonic impedance is an efficient tool for assessing the state of distribution networks containing harmonics. The unfavorable operating conditions are anticipated based on the network harmonic impedance values, and the means of intervention are selected. Harmonic impedance monitoring and using it in expert systems for operating condition optimization will increase in the future. Power factor correction by shunt capacitor switching in electrical networks containing harmonics can lead to harmonics amplifications by harmonic voltage increasing and capacitors thermal overstressing by great values of the currents flowing through them. This chapter proposes a method for practical determination of harmonic impedance. Based on its values, a quick method is developed to anticipate the harmonic voltages and current amplifications that can occur when a shunt capacitor is installed for power factor correction. Amplification factors are calculated depending on the equivalent harmonic impedance of the network seen in the compensation bus. A distribution network containing harmonics is modeled using MatLab Simulink, and harmonic impedance is determined by simulation in different operating conditions. Using the values of the harmonic impedance and the capacitive reactance of the capacitor bank that is connected for power factor correction, the amplification of the harmonic voltages and currents is estimated by calculus. The results obtained by calculus are then compared with the values obtained by simulation after the connection of the capacitor bank to the network. In conclusion, the chapter proves that the network harmonic impedance is a useful tool to estimate the harmonics amplification caused by power factor correction using shunt capacitor banks.

Keywords: distribution electrical networks, reactive power compensation, capacitor banks, harmonic impedance, harmonics amplification

---

1. Introduction

Harmonics are making their presence felt in electrical distribution networks due to both the use of nonlinear devices in the consumers' area and the high development of distributed generation [1, 2].

The harmonics of the electrical networks have negative effects like increasing active power losses, resonance over voltages, increasing the potential of the transformers' neutral point, over currents in three-phase networks which can lead to difficulties regarding the protection relay selectivity, and capacitor banks overstressing. The effects of some types of loads like office equipment, air conditioning units, lighting devices, motor drives, and household equipment are presented in [3]. This study discusses the influence of capacitor banks on the network harmonics conditions.

Connecting a capacitor bank to the electrical network leads to the resonance phenomena, which, superimposed over the harmonic current flow on frequencies close to the resonance frequency, result in high harmonic voltage values at the capacitor terminals and high harmonic currents flowing through the capacitor bank. Over voltages lead to insulation overstressing, and high currents produce a capacitor bank heating that is also reflected in insulation damage [4, 5].

There are many methods for network harmonic impedance determination, which are presented in many literature works [6–12]. Even the operation of switching the capacitor bank can be used for harmonic impedance measuring [13], and online determination of the harmonic impedance is the subject of many works [14, 15]. One problem that is less noticed in these works is how to determine the phase of the complex harmonic impedance.

This chapter presents a method, which uses the network harmonic impedance like a tool to anticipate the harmonic condition amplifications in the case of reactive power compensation by capacitor banks.

In order to reach this objective, Section 2 presents a method to determinate the complex harmonic impedance. In order to find the phase of the impedance, the series–parallel circuit transformation is used. A mathematical model to anticipate the harmonic currents and voltage amplification, based on the harmonic impedance values, when a capacitor bank is connected to the network, is also described in Section 2. The mathematical model is validated, in Section 3, by a MatLab Simulink simulation for an electrical distribution network that comprises a bus with a capacitor bank installed for power factor correction. The results obtained by simulation are compared with those obtained by calculus from the mathematical model. Section 4 presents, step by step, how the distribution network is implemented on a Matlab Simulink model. Section 5 comprises the conclusions of the chapter.

2. Amplifying harmonic conditions assessment

2.1. Determination of the network harmonic impedance

The simplest method for harmonic impedance measurement supposes the changing of network status. The disturbance can be created by switching a network component in the bus where the network harmonic impedance must be measured. The main steps of this method are as follows [16]:

1. Voltage and current waveforms recording before the disturbance.

2. Changes are then made to the status of the network.

3. Voltage and current waveforms recording after the disturbance.

4. Network harmonic impedances determination:

$$\underline{Z}_k = \frac{\underline{U}_{k-\text{pre}} - \underline{U}_{k-\text{post}}}{\underline{I}_{k-\text{pre}} - \underline{I}_{k-\text{post}}} \tag{1}$$

where $\underline{U}_{k-\text{pre}}$ and $\underline{I}_{k-\text{pre}}$ are the pre-disturbance harmonic voltage and current, and $\underline{U}_{k-\text{post}}$ and $\underline{I}_{k-\text{post}}$ are the post-disturbance harmonic voltage and current.

The impedance $\underline{Z}_k$ does not include the switched load.

Because the harmonic impedance determined using the expression (1) is parallel impedance, to obtain the phase of the impedance, we have to transform the parallel schema into series schema.

The series harmonic impedance can be determined using the following expression:

$$\underline{Z}_{kS} = R_S + j \cdot X_{kS} \tag{2}$$

The parallel harmonic impedance can be determined using the following expression:

$$\frac{1}{\underline{Z}_{kp}} = \frac{1}{R_p} + \frac{1}{jX_{kp}} \tag{3}$$

$$\underline{Z}_{kp} = \frac{1}{\frac{jX_{kp}+R_p}{jX_{kp}\cdot R_p}} = \frac{jX_{kp}\cdot R_p}{jX_{kp} + R_p} \cdot \frac{R_p - jX_{kp}}{R_p - jX_{kp}} = j \cdot \frac{R_p^2 \cdot X_{kp}}{R_p^2 + X_{kp}^2} + \frac{R_p \cdot X_{kp}^2}{R_p^2 + X_{kp}^2} \tag{4}$$

Knowing that:

$$\underline{Z}_{kp} = \underline{Z}_{kS} = R_S + j \cdot X_{kS} \tag{5}$$

Results:

$$R_S = \frac{R_p \cdot X_{kp}^2}{R_p^2 + X_{kp}^2}$$

$$X_{kS} = \frac{R_p^2 \cdot X_{kp}}{R_p^2 + X_{kp}^2}$$

(6)

We write now these parameters of the series impedance depending on the absolute value of the harmonic impedance, which can be practically determined using (1).

$$Z_{kp}^2 = Z_{kS}^2 = R_S^2 + X_{kS}^2 = \frac{R_p^2 \cdot X_{kp}^2}{R_p^2 + X_{kp}^2}$$

(7)

Using expressions (6) and (7), we can write the series resistance as follows:

$$R_S = \frac{R_p \cdot X_{kp}^2}{R_p^2 + X_{kp}^2} \cdot \frac{R_p}{R_p} = \frac{Z_{kp}^2}{R_p}$$

(8)

The value of the parallel resistance, R_p, from the expression (8) can also be practically determined; it is actually the value of the harmonic impedance corresponding to the resonance frequency. Using the expression (7), the value of the series reactance can now be determined:

$$X_{kS} = \sqrt{Z_{kS}^2 - R_S^2}$$

(9)

With the values of the series parameters of the harmonic impedance, we can now determine the phase of the complex harmonic impedance:

$$\varphi_k = arctg\left(\frac{X_{kS}}{R_S}\right)$$

(10)

2.2. Mathematical model of amplifying the harmonic conditions by reactive power compensation

Switching shunt equivalent impedance in a network containing harmonics, conduce to more or less influence on the harmonic conditions, depending on the character and the value of the impedance.

Switching a shunt capacitor bank in order to improve the power factor in a bus of a distribution network containing harmonics can conduce to the amplification of the harmonic conditions. This is possible only when the harmonic currents are flowing in the network, with the frequencies close to the parallel resonance frequency. The resonance frequency occurs between the capacitance of the capacitor bank and the equivalent inductance of the network.

Figure 1 presents the equivalent schema for the k range harmonic, for the cases before and after the shunt capacitor impedance switching ($\underline{Z}_{kCB}$) in the point of common coupling (PCC).

The notations used in **Figure 1** are:

$\underline{I_k}$ is the harmonic current provided by an equivalent fictive current source. It corresponds to all the k range harmonic current sources existing in the network, and it produces the harmonic impedance corresponding to PCC, $\underline{Z_{knet}}$, and the harmonic voltage drop $\underline{U_k}$.

The harmonic equivalent current source for k range ($\underline{I_k}$) does not change its value after connecting the capacitor bank (CB) represented by the equivalent reactance, $\underline{Z_{kCB}}$.

In order to determinate the mathematical expressions that will be used, the harmonic imped-ance "seen" in the *PCC* before the installation of $\underline{Z_{kCB}}$ is separated in two components:

$\underline{Z_{kload}}$ is the transversal impedance of the load existing before the disturbance, usually a load impedance and

$\underline{Z_{knet}}$ is the equivalent impedance of the rest of the network.

The harmonic voltage of k range in PCC is:

$$\underline{U_k} = \underline{I_k} \cdot \underline{Z_k} = \underline{I_{kload}} \cdot \underline{Z_{kload}} \tag{11}$$

in which: $\underline{Z_k} = \underline{Z_{knet}} \| \underline{Z_{kload}} = \frac{\underline{Z_{knet}} \cdot \underline{Z_{kload}}}{\underline{Z_{knet}} + \underline{Z_{kload}}}$.

Results of the component of the harmonic current distributed through $\underline{Z_{kload}}$:

$$\underline{I_{kload}} = \underline{d_{load}} \cdot \underline{I_k} \tag{12}$$

where

$$\underline{d_{kload}} = \frac{\underline{Z_k}}{\underline{Z_{kload}}} \tag{13}$$

We can also write:

$$\underline{I_k} = \underline{I_{kload}} \cdot \frac{\underline{Z_{kload}}}{\underline{Z_k}} \tag{14}$$

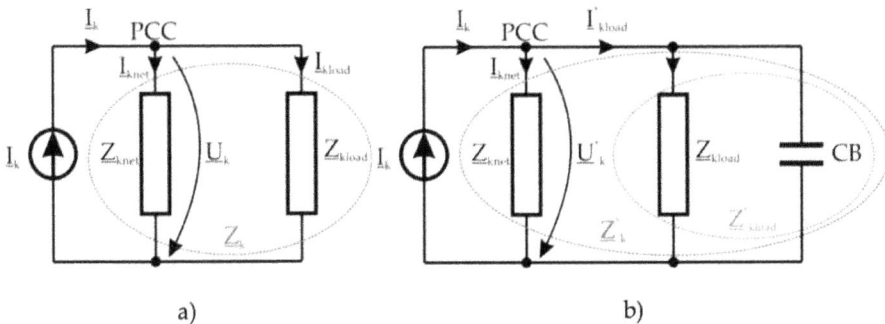

Figure 1. The equivalent circuit: (a) before the capacitor connection; (b) after the capacitor connection.

After the shunt capacitor bank connecting, Z_{kCB}, supposing that the harmonic current sources and the harmonic impedance of the network do not change, we can write:

$$U_k^* = I_k \cdot Z_k^* = I_{kload}^* \cdot Z_{kload}^* \tag{15}$$

Consulting expressions for the harmonic impedance corresponding to PCC after the reactive power compensation (Z_{kCB} switching):

$$Z_k^* = Z_k \| Z_{kCB} = \frac{Z_k \cdot Z_{kCB}}{Z_k + Z_{kCB}} \tag{16}$$

The total transversal impedance:

$$Z_{kload}^* = Z_{kload} \| Z_{kCB} = \frac{Z_{kload} \cdot Z_{kCB}}{Z_{kload} + Z_{kCB}} \tag{17}$$

From (15), we have the following result:

$$I_{kload}^* = I_k \cdot \frac{Z_k^*}{Z_{kload}^*} \tag{18}$$

Using the expression of I_k for the regimen before the reactive power compensation, Z_{kCB} (expression (14)):

$$I_{kload}^* = I_{kload} \cdot \frac{Z_{kload}}{Z_k} \cdot \frac{Z_k^*}{Z_{kload}^*} \tag{19}$$

or

$$I_{kload}^* = E_{I_{load}} \cdot I_{kload} \tag{20}$$

where the notation $E_{I_{load}}$ represents the load current amplification factor, pursuant to the reactive power compensation, Z_{kCB}:

$$E_{kI_{load}} = \frac{Z_{kload}}{Z_k} \cdot \frac{Z_k^*}{Z_{kload}^*} \tag{21}$$

Using expressions (16) and (17) in (21), we obtain:

$$E_{kI_{load}} = \frac{Z_{kload} + Z_{kCB}}{Z_k + Z_{kCB}} \tag{22}$$

So, knowing the impedances Z_k and Z_{kCB} from the initial regimen, and the initial load harmonic current (I_{kload}), we can find the final load current I_{kload}^* and also the harmonic voltage in PCC after the capacitor bank connection Z_{kCB}:

$$\underline{U}_k^* = \underline{I}_{kload}^* \cdot \underline{Z}_{kload}^* \tag{23}$$

The harmonic voltage amplification factor in PCC is calculated by the following expression:

$$\underline{F}_{U_k} = \frac{\underline{U}_k^*}{\underline{U}_k} = \frac{\underline{Z}_k^*}{\underline{Z}_k} = \frac{\frac{\underline{Z}_k \cdot \underline{Z}_{kCB}}{\underline{Z}_k + \underline{Z}_{kCB}}}{\underline{Z}_k} = \frac{\underline{Z}_{kCB}}{\underline{Z}_k + \underline{Z}_{kCB}} \tag{24}$$

and

$$F_{U_k} = \frac{|\underline{U}_k^*|}{|\underline{U}_k|} = \frac{|\underline{Z}_{kCB}|}{|\underline{Z}_k + \underline{Z}_{kCB}|} \tag{25}$$

The total harmonic distortion (THD) of the PCC voltage after connecting to the capacitor bank must be lower than the highest value:

$$THD_{U^*} = \frac{\sqrt{\sum_{k=2}^{n} U_k^{*2}}}{U_1} = \frac{\sqrt{\sum_{k=2}^{n} (F_{U_k} \cdot U_k)^2}}{U_1} \leq THD_{Umax} \tag{26}$$

In addition, the voltage in PCC must be lower than the highest value imposed by the capacitor bank manufacturer, usually 10% over the nominal voltage.

Knowing the voltage bus in PCC and the capacitor bank reactance, we can also check if the current trough in the capacitor bank is lower than the highest value imposed by the capacitor bank manufacturer, usually 30% over the nominal current:

$$I_{CB} = \sqrt{\sum_{k=1}^{n} I_{kCB}^2} \leq I_{CBmax} \tag{27}$$

3. Simulation results, discussions

In order to estimate the changes regarding the harmonic conditions in a bus of a network, due to the installation of transversal impedance, it means to determinate the new harmonic voltages and harmonic currents, using the initial state of the network.

The validation of the mathematical model was done on a distribution electrical network comprising three voltage levels: 20, 6, and 0.4 kV, respectively. The configuration of the network and the main characteristics of its elements are presented in **Figure 2**.

To simulate the permanent normal operating conditions, the MatLab Simulink soft is used.

The harmonic regimen was obtained using harmonic current sources for the harmonics 5, 7, and 13, in three of the busses of the network (noted 1, 2, 3)—the load busses. These harmonic sources model the nonlinear components of the loads.

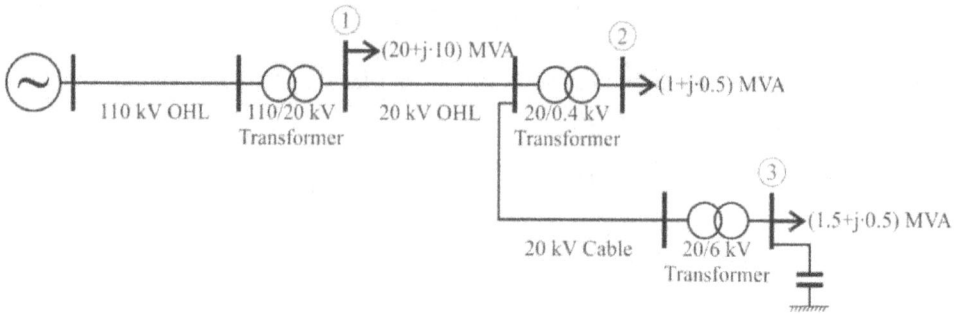

Figure 2. The electrical distribution network used to validate the mathematical model.

For the validation of the mathematical model, bus no. 3 was selected.

The transversal impedance that will be installed has a capacitive character, and it corresponds to 0.3 MVAr capacitive reactive power on the fundamental frequency. The installation of a CB for power factor correction or the voltage control is an operation with a high impact for both the network and the capacitors. If the network contains harmonics, the effect of connecting a capacitor bank will be harmonic conditions amplification and capacitors thermal overstressing risk.

The ampleness of both effects can be anticipated using the state values of the network before the installation like it was proved earlier.

The amounts obtained by MatLab simulation before the installation of the CB are presented in **Table 1**. The notation represents the equivalent impedance of the linear component of the load.

By the successively installation of a CB having 0.3 MVAr capacitive reactive power, the values obtained using the method presented in this chapter, respectively, by MatLab simulation, are written in **Tables 2** and **3**.

The simulation was deliberately done for a case with high harmonic conditions—see the values in **Table 1**. It results in $THD_U = 20\%$ for the bus 3, before the reactive power compensation. For the bus 3, where is connected also a load, for the current absorbed by this, we have $THD_I = 19\%$. We only concentrate on the linear component of the load connected in bus 3. From **Table 3**, we observe that after the reactive power compensation, the voltage THD in the

k	$\underline{Z}_k$		$\underline{Z}_{k\ load}$		U_k (V)	THD_U	$I_{k\ load}$ (A)
	$|\underline{Z}_k|$ (Ω)	Phase (deg)	$|\underline{Z}_{k\ load}|$ (Ω)	Phase (°)			
1	1.633	51.06	22.76	18.43	3182	0.2	15.75
5	6.376	62.21	23.946	86.18	430.34		13.92
7	8.542	57.2	23.972	87.27	524.09		14.00
13	13.58	39.74	23.992	88.53	376.01		17.75

Table 1. The values from steady-state operation before the capacitor bank switching.

k	F_{U_k}	U_k^* (V)	THD_{U^*}	$F_{kI_{load}}$	I_{kload}^* (A)	$THD_{I^*_{load}}$	I_{kCAP}^* (A)
1	1.01	3215.9	0.241	0.967	135.1	0.457	26.799
5	1.29	430.27		1.76	24.52		17.928
7	1.56	524.06		2.622	36.72		30.5708
13	0.88	376.16		2.43	43.27		40.7543

Table 2. The values calculated after the capacitor bank switching.

k	U_k^* (V)	THD_{U^*}	I_{kload}^* (A)	$THD_{I^*_{load}}$	I_{kCAP}^* (A)
1	3216.0	0.2411	135.18	0.4573	26.8
5	430.31		24.526		17.93
7	524.09		36.728		30.57
13	376.01		43.268		40.73

Table 3. The values obtained by MatLab simulation after the capacitor bankswitching.

bus 3 will increase to 24.11%. This is due to the high distortion of the current that will flow through the capacitive impedance after its installation.

Comparing now the values obtained by calculus (**Table 2**) with the values measured in MatLab (**Table 3**), we observe a good proximity.

4. MatLab implementation of the proposed model

MatLab Simulink simulation software is frequently used in a large area of engineering fields like communications, electrical engineering, mechanical engineering, digital image processing, and others, both for research and education [17, 18].

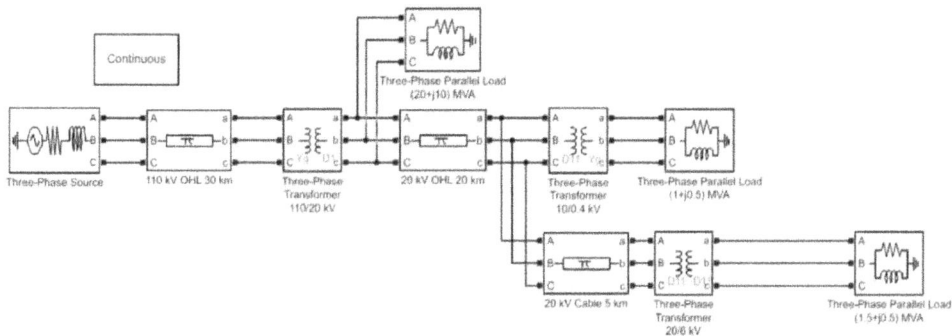

Figure 3. The electrical distribution network implemented in MatLab Simulink before the reactive power compensation.

In this chapter, the proposed mathematical model was validated using Matlab Simulink. The electrical network presented in **Figure 2** was modeled in Simulink, **Figure 3**, as follows:

The electrical system was modeled using a 110-kV three-phase balanced source having Y grounded connection.

The electrical overhead lines (OHL) and cables are modeled using three-phase PI section lines. The line parameters resistor (R), inductor (L), and capacitor (C) are specified as positive- and zero-sequence parameters that take into account the inductive and capacitive couplings between the three-phase conductors, as well as the ground parameters. This method of specifying line parameters assumes that the three phases are balanced.

A three-phase two windings transformer model is used for all the transformers of the network. The connection of the windings is Y grounded for high voltage (110 kV) and D for medium voltage (20 and 6 kV). The parameters can be written in pu units or SI units. Changing the Units parameter from pu to SI, or from SI to pu, automatically converts the parameters displayed in the mask of the block.

The loads are modeled using parallel R L C circuits defined by active and reactive powers. Only the elements corresponding to non-zero powers are displayed. In this example, the reactive power of the loads is inductive reactive power.

This model is used for the steady-state operation condition before the capacitor bank switching.

Figure 4 presents the MATLAB Simulink model of the electrical distribution network with the capacitor bank for power factor correction connected in bus No.3. The capacitor bank is modeled as a capacitive load defined by the capacitive reactive power.

A detail of the compensation bus is presented in **Figure 5**. Here, the harmonic current sources for the 5th, 7th, and 13th harmonics are presented. These are ideal current sources having the following parameters: peak amplitude, phase, and frequency. A three-phase measurement

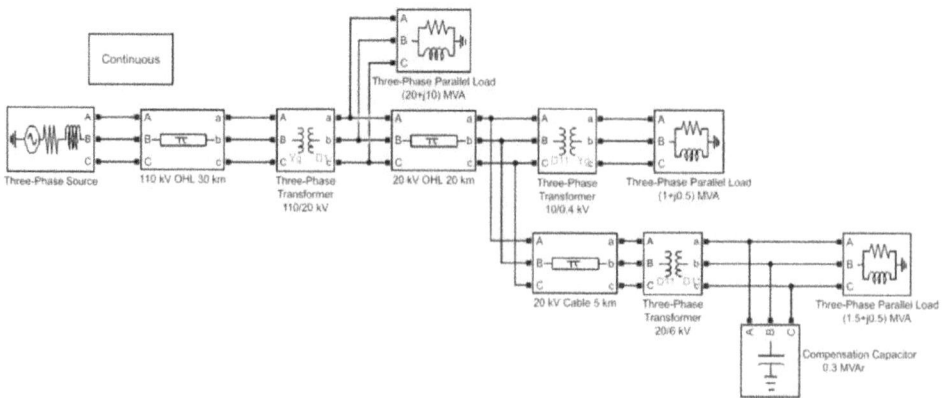

Figure 4. The electrical distribution network implemented in MatLab Simulink after the reactive power compensation.

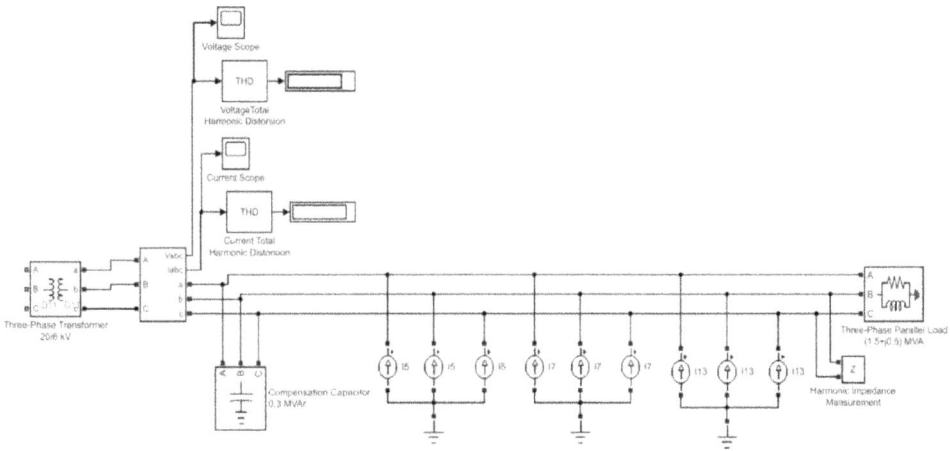

Figure 5. Details of the compensation bus.

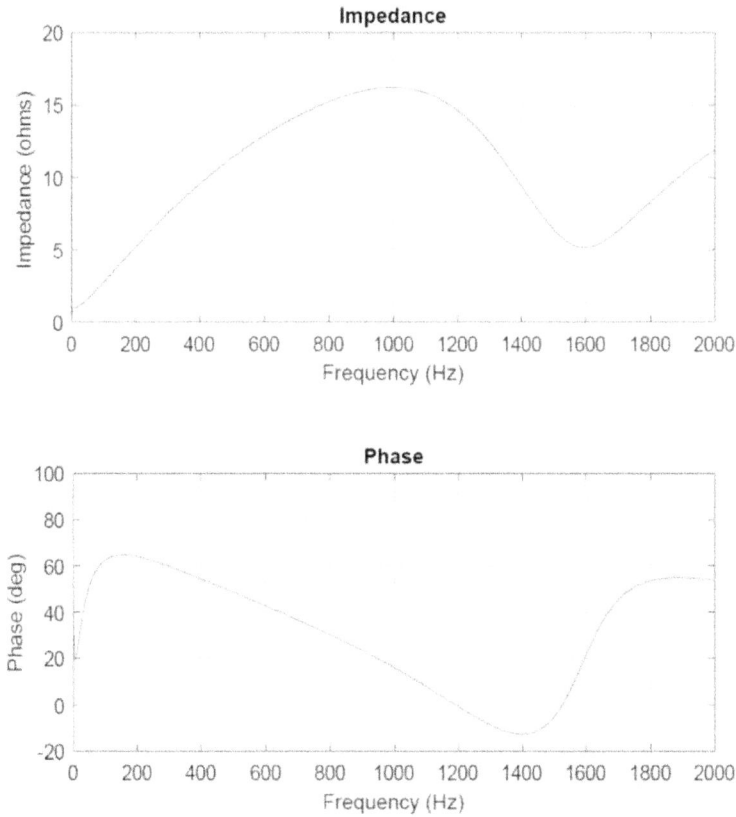

Figure 6. Harmonic impedance before the compensation.

block for both voltages and currents are also inserted in the model. The user can choose phase-to-phase or phase-to-ground measuring. Knowing the harmonic currents and harmonic voltages, the total harmonic distortion (THD) factor both for voltage and current can be calculated.

The harmonic impedance of the network is measured using a dedicated block named "Impedance Measurement" which measures the impedance in a bus of the network as a function of the frequency (harmonic impedance). The harmonic impedance (magnitude and phase) is displayed by using the appropriate tool from the Powergui block. The results obtained in the two cases are presented in **Figures 6** and **7**.

In order to highlight the influence of the capacitor reactive power amounts on the values of the harmonic impedance, the following compensation steps are considered: 0.2, 0.3, and 0.4 MVAr. Corresponding to these steps of the capacitor reactive power, the resonance frequency is much lower, while the reactive power increases: 404 Hz for 0.4 MVAr, 460 Hz for 0.3 MVAr, and 550 Hz for 0.2 MVAr. The results are presented in **Figure 8**.

The influence of the load active power is also interesting to study. For a constant reactive power of compensation (0.3 MVar), the resonance frequency does not change, but the

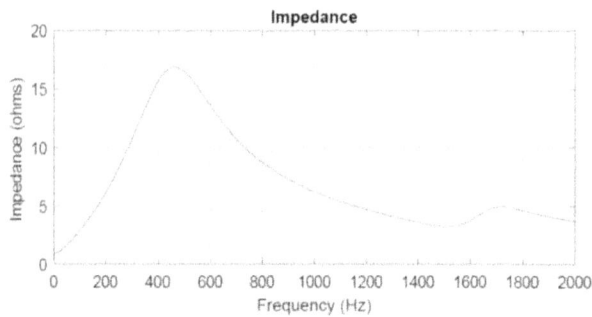

Figure 7. Harmonic impedance after the compensation.

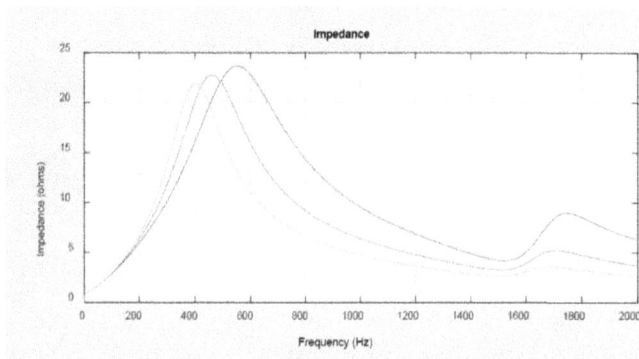

Figure 8. The influence of the amount of reactive power compensation.

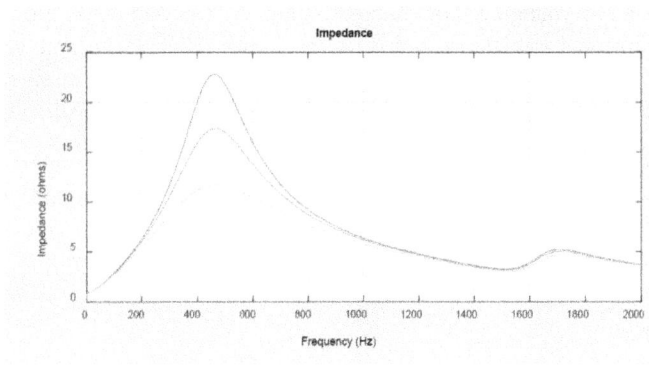

Figure 9. The influence of the amount of load active power.

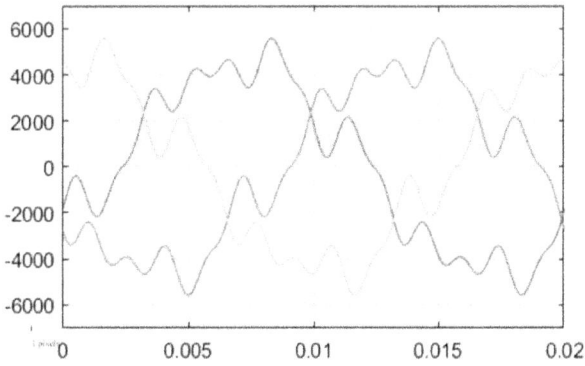

Figure 10. Voltages waveforms in bus 3 after the capacitor bank switching.

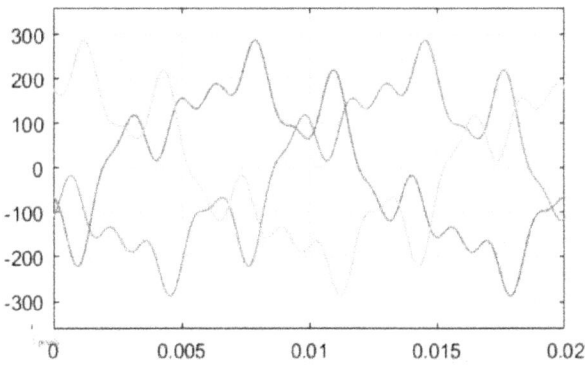

Figure 11. Currents waveforms in bus 3 after the capacitor bank switching.

amplification factor is changing. For different values of the load active power (1, 1.5, and 2.5 MW), the harmonic impedance has different values (22.8, 17.32, and 11.7 Ω) and the resonance frequency has a constant value: 460 Hz. The results are presented in **Figure 9**. Changes of the load reactive power do not influence the results of the harmonic impedance.

Scopes are installed for both voltage and current. The results for time-domain simulation for one cycle (0.02 s), after switching the capacitor bank, are presented in **Figure 10** for voltages and in **Figure 11** for currents.

5. Conclusions

The scope of this chapter consists in highlighting the importance of harmonic impedance study in a bus of an electrical network with harmonics where a capacitor bank will be installed. Capacitor bank switching leads to an amplification of the harmonic impedance for the frequencies close to the resonance frequency.

The harmonic voltage amplification will lead to high values of voltage to the capacitor terminals and the effect will be an electrical overstressing of the capacitor banks. Harmonic current amplification will lead to high values of current trough the capacitor bank and the effect will be a thermic overstressing of the capacitor banks.

A mathematical model for harmonic impedance determination is presented, and, based on its values, anticipation of harmonic amplification is proved. A numerical example is presented for a distribution network containing harmonics and capacitor banks for power factor correction.

The validation of the mathematical model was done by MatLab Simulink simulation. The presented electrical distribution network is modeled using elements from the Library Browser like: three-phase balanced source for electrical system modeling, PI section lines for overhead lines and cables, three-phase two windings transformers, and three-phase loads. Harmonic conditions are simulated using ideal current sources for different frequencies. Three-phase measurement blocks are used for measuring harmonic voltages and currents, and harmonic impedance is also measured using a dedicated block.

We observe a good proximity of the results obtained by calculus and by simulation, MatLab Simulink being a useful tool for research and education in the field of power engineering.

Author details

Alexandru Băloi* and Adrian Pană

*Address all correspondence to: alexandru.baloi@upt.ro

"Politehnica" University of Timisoara, Timisoara, Romania

References

[1] Sikorski T, Rezmer J. Distributed generation and its impact on power quality in low-voltage distribution networks. In: Luszcz J, editor. Power Quality Issues in Distributed Generation. Rijeka: InTech; 2015. DOI: 10.5772/61172 Available from: https://www.intechopen.com/books/power-quality-issues-in-distributed-generation/distributed-generation-and-its-impact-on-power-quality-in-low-voltage-distribution-networks

[2] Canova A, Giaccone L, Spertino F, Tartaglia M. Electrical impact of photovoltaic plant in distributed network. IEEE Transactions on Industry Applications. 2009;**45**:341-347

[3] Kocatepe C, Yumurtacı R, Arıkan O, Baysal M, Kekezoğlu B, Bozkurt A, Fadıl Kumru C. Harmonic effects of power system loads: An experimental study. In: Zobaa DA, editor. Power Quality Issues. Rijeka: InTech; 2013. DOI: 10.5772/53108 Available from: https://www.intechopen.com/books/power-quality-issues/harmonic-effects-of-power-system-loads-an-experimental-study

[4] Farag AS, Wang C, Cheng TC, Zheng G, Du Y, Hu L, Palk B, Moon M. Failure analysis of composite dielectric of power capacitors in distribution systems. IEEE Transactions on Dielectrics and Electrical Insulation. 1998;**5**(4)

[5] Boonseng C, Chompoo-Inwai C, Kinnares V, Nakawiwat K, Apiratikul P. Failure analysis of dielectric of low voltage power capacitors due to related harmonic resonance effects. In: IEEE Power Engineering Society Winter Meeting; Columbus OH; 2001

[6] Borkowski D, Wetula A, Bien A. New method for noninvasive measurement of utility harmonic impedance. In: Proceedings of Pes General Meeting; San Diego; 2012

[7] Hui J, Freitas W, Vieira JCM, Yang H, Liu Y. Utility harmonic impedance measurement based on data selection. IEEE Transactions on Power Delivery. October 2012;**27**(4):2193-2202

[8] Karimzadeh F, Esmaeili S, Hosseinian SH. A novel method for noninvasive estimation of utility harmonic impedance based on complex independent component analysis. IEEE Transactions on Power Delivery. August 2015;**30**(4):1843-1852

[9] Shusen LI, Xiao Y, Cheng J. Dual synchronization incremental method for harmonic impedance measurement. In: IEEE 15th International Conference on Harmonics and Quality of Power; 2012; Hong Kong. pp. 433-437

[10] Asiminoaei L, Teodorescu R, Blaabjerg F, Borup U. A digital controlled PV-inverter with grid impedance estimation for ENS detection. IEEE Transactions on Power Electronics. November 2005;**20**(6):1480-1490

[11] Stiegler R, Meyer J, Schegner P, Chakravorty D. Measurement of network harmonic impedance in presence of electronic equipment. In: IEEE International Workshop on Applied Measurements for Power Systems (AMPS); 2015; pp. 49-54

[12] Hui J, Yang H, Lin S, Ye M. Assessing utility harmonic impedance based on the covariance characteristic of random vectors. IEEE Transactions on Power Delivery. 2010;25:1778-1786

[13] Nan W, Xiaoming Z. Capacitor-switching based method of network harmonic impedance measurement via TLS-ESPRIT algorithm. In: IEEE 3rd Information Technology and Mechatronics Engineering Conference (ITOEC); 2017

[14] Serfontein D, Rens J, Botha G, Desmet J. Continuous harmonic impedance assessment using online measurements. In: IEEE International Workshop on Applied Measurements for Power Systems (AMPS); 2015. pp. 55-60

[15] Baloi A, Pana A, Molnar-Matei F. Contributions on harmonic impedance monitoring in smart grids using virtual instruments. In: 2nd IEEE PES International Conference and Exhibition on Innovative Smart Grid Technologies; 2011; pp. 1-5

[16] Wilsun X, Ahmed EE, Zhang X, Liu X. Measurement of network harmonic impedances: Practical implementation issues and their solutions. IEEE Transactions on Power Delivery. 2002;17(1):210-216

[17] Valdman J, editor. Applications from Engineering with MATLAB Concepts. Rijeka: InTech; 2016. p. 284. DOI: 10.5772/61386

[18] Bennett K, editor. MATLAB Applications for the Practical Engineer. Rijeka: InTech; 2014. p. 664. DOI: 10.5772/57070

Dynamic Modeling for Open- and Closed-loop Control of PMSG based WECS with Fuzzy Logic Controllers

Anbarasi Jebaselvi Jeya Gnanaiah David and
Meenakshi Veerappan

Additional information is available at the end of the chapter

http://dx.doi.org/10.5772/intechopen.72693

Abstract

The high risk in developing a more advanced wind power generator with scientific and technological know-how, heavy loss in maintaining the accessories of a wind plant and stochastic nature of wind energy make the maximum energy retrieval questionable, but still optimum wind energy extraction can be achieved by operating the wind turbine generator (WTG) in a variable-speed, variable-frequency mode with different types of wind electric generators (WEGs). In this chapter, maximum power from wind using permanent magnet synchronous generator (PMSG) is made possible by using intelligent controllers, namely fuzzy logic controllers. The chapter also discusses the simulated results obtained from modeling, simulation, and analysis of this PMSG-based wind energy conversion system (WECS) for both open- and closed-loop control strategies. PMSG suffers drastically from load and strong decay of magnetic field, which tends to reduce the generated voltage at the stator terminals, making it difficult for isolated operation and thus the whole analysis is done with grid-connected network. The other major limitations include loss of flexibility in field flux control, hence intelligent techniques like fuzzy logic mechanism are attempted along with space-vector modulation (SVM) to have a smooth control of field flux and load power management in PMSG.

Keywords: PMSG, FLC, SVM, WECS, WTG, power quality

1. Introduction

Nowadays, though there is an enormous degree of power production and energy conservation, there is a void which could be filled by familiarizing renewables into the power market. The major renewable energy sources include wind, solar, fuel cells, geo-thermal, tidal waves, and biofuel systems; by analyzing the pros and cons of all these sources, wind and solar energy seem to be economically accessible and less prone to pollution and thus becomes inevitable. Gearless construction with a total elimination of DC excitation system, compactness in yielding maximum power extraction, smooth grid interfacing, and ease in handling fault ride make PMSG the most sought appliance in wind power industry.

Since wind speed and wind power associated with wind force is not constant, the torque developed is not constant as well. Therefore, whenever PMSG is coupled with wind turbine, the output voltage generated varies in both frequency and magnitude. In order to synchronize load bus with grid of fixed frequency AC, the power output of PMSG is converted to DC and then inverted to 50 Hz AC. In addition, the change in DC-link voltage fluctuates in an uncontrolled manner, which has to be regulated by suitable modulation strategy applied to the inverter on grid side.

Considering these facts, a summary of the literature survey has been included to compare the research outcomes carried out so far on PMSG-based WECS. Various control strategies along with the modeling of PMSG were analyzed in order to control DC-link voltage, output power and pitch of the turbine in these articles elaborately [1–4]. Several maximum power point tracking (MPPT) algorithms were used to obtain the maximum power from the high-power turbines [5–9] with a simple and effective controller to ensure smooth control over voltage, frequency, and power output. The performance and analysis of an ultra-large wind turbine using validated models of mechanical and electrical systems of a wind turbine have been conducted under various conditions of step changes in wind speed generated by TurbSim [10]. A multipole PMSG model with maximum power point tracking (MPPT) mechanism was developed to extract maximum power in this paper [11]. Vector control and sliding mode control techniques were implemented to maximize the electromagnetic torque to regulate the DC bus voltage and concluded that PMSG was the best for wind power generation systems with good performance characteristics of speed, flux, and torque [12].

Various control algorithms were proposed and implemented to control the speed of PMSG with respect to wind speed in both generation and grid side to maintain the power flow [13–17]. PMSG performed well for step changes of load while implementing direct torque control (DTC)-based space-vector modulation, and the regulation of power factor was quiet good [18, 19]. Sensorless control strategies were compared with various modeling techniques, and simulation results were analyzed in these articles [20–23].

The two modulation techniques, namely sinusoidal pulse width modulation (SPWM) and space-vector pulse width modulation technique (SVPWM), are compared, and the resultant voltage space vector is found to be rotating consistently at synchronous speed with a magnitude 1.5 times greater than the peak value of the phase voltage [24]. The maximum power is obtained using the SVPWM modulation technique with the DC-link voltage kept at stable level to obtain decoupled control of active and reactive power. The DC voltage utilization ratio is around 71% of the DC-link voltage as compared to the conventional sinusoidal pulse width modulation which is found to be 61.2%. Space-vector PWM generates less harmonic distortion in the output voltage/current waveform in comparison with a sine PWM [25].

In this paper [26], the simulated results show that the proposed Fuzzy-PI controller is very effective in improving the transient stability of overall wind farm systems during temporary and permanent fault conditions. The main advantage of the proposed MPPT method in [27] is that there is no need of measuring wind velocity and generator speed. As such, the control algorithm is independent of turbine characteristics, achieving the fast dynamic responses with nonlinear fuzzy logic systems.

With the use of PI controller, the harmonic spectrum (THD) for load voltage and load current are found to be 1.31 and 37.92%, respectively. The harmonic spectrums for load voltage and load currents with fuzzy logic are 0.47 and 35.72%, respectively. Simulation results and harmonic spectrum obtained in [28] demonstrate that the fuzzy-based controller works very well and shows very good dynamic and steady-state performance. The research contributions of this technical article [29] deals with the development of single-ended primary-inductor converter (SEPIC)-based FLC-WECS which can maintain constant voltage at the output with minimized ripple content and improvement of dynamic response using fuzzy logic controller.

This paper [30] focuses on robust control based on T-S approach that allows tracking of rotational speed, stator current, and voltage references which correspond to the optimum power and therefore operate with the maximum power.

Here, this chapter discusses open-loop as well as closed-loop method with intelligent controller, and it is confirmed that sufficient revenue could be generated from it by implementing the same in real time with very little computation. The following sections dealt in detail about modeling of PMSG based WECS with its simulation, open-loop and closed-loop control with fuzzy logic technique and concluded with the note that the generated power exceeds load power, hence apart from meeting the load demands, it supplies the battery load in addition.

2. Modeling and simulation of PMSG-based WECS

The configuration and control schematic of PMSG-based WECS are depicted in **Figure 1**. The captured kinetic energy of wind rotates the generator rotor and cuts the magnetic field set by the stator and thus produces electric power. The generated AC output voltage is converted into DC using bridge rectifier, and again back to AC by means of DC/AC inverter. A DC-link voltage is connected in between the generator and grid-side converters/inverters.

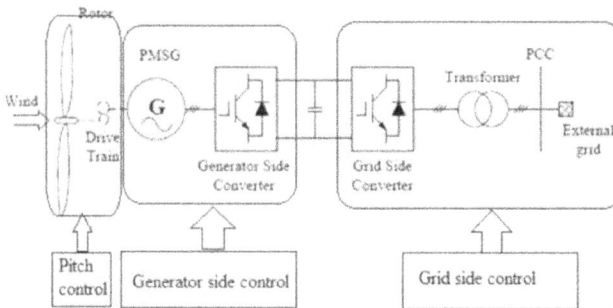

Figure 1. Configuration and control logic used in PMSG-based WECS.

Voltage-fed, current-controlled inverter with insulated-gate bipolar transistor (IGBT) as a switching device is used to control and synchronize both the stator output voltage and frequency with the grid components. The stator voltage is stepped up by a transformer and passed to the network through a point of common coupling (PCC) where the injection of wind power has an impact on voltage magnitude, its flicker, and its output waveform. The magnetic core losses are minimized by the inclusion of a transformer to increase the overall efficiency.

This section presents several modules like model of PMSG, a model of rectifier, a model of boost converter, and a voltage-fed inverter model with SVPWM along with voltage vectors. The major system variables continuously monitored and controlled generate optimal power at different wind speeds by PMSG and thus active and reactive power are injected into the grid and to the DC bus.

The control of rectifiers and converters is done by sinusoidal PWM control technique at the generator-side control, while the grid-side inverter control includes SVM technique. The dynamic model of PMSG with its rotor speed control system is represented by the equations of generator in a reference coordinates system rotating synchronously with the magnetic flux. Since the stator current vector is represented by rotor flux with respect to d-q axis reference system, i_d, i_q and the electromagnetic torque are related with each other through this vector. The magnetic axis of the rotor is fixed as reference for spatial orientation of fictitious rotor windings.

Considering the inductances L_{ds} and L_{qs} in the stator of the generator which are equal along the direct and quadrature axes, respectively and $L_{ds} = L_{qs} = L_s$ at steady-state condition, the stator equations in terms of d and q axes are given in Eqs. (1) and (2) as

$$\frac{di_{ds}}{dt} = \frac{1}{L_{ds} + L_{ls}}(-R_s i_{ds}) + \omega_e (L_{qs} + L_{ls}) i_{qs} + u_d \tag{1}$$

$$\frac{di_{qs}}{dt} = \frac{1}{L_{qs} + L_{ls}}(-R_s i_{sq}) - \omega_e \left((L_{ds} + L_{ls}) i_{ds} + \frac{d\psi}{dt} ds\right) + u_q \tag{2}$$

where L_{ls} is the leakage inductance of the stator referred to d and q axes.

PMSG is modeled by using derived mathematical equations and simulated using SimPower Systems library available in MATLAB/Simulink. The simulation is carried out for a grid-connected PMSG in both open- and closed-control modes. Each and every individual model is integrated to analyze the complete system behavior. Since wind power is unreliable, the PMSG output is not stable. Hence, to synchronize the generated output voltage with the inverter frequency, PLL is used. The generator is directly connected to the grid through a full-scale back-to-back power converter.

The power converter decouples the generator and the grid. In addition, this full-scale power converter allows full controllability of the entire system. Both the generator/machine and grid-side converters operate in rectifier or inverter mode and thus maintain the bidirectional power flow. The generated three-phase AC voltage is stepped up and passed through the utility grid. The generator/stator-side converter mainly controls the speed of generator to obtain the maximum power output even at low wind speeds. The grid-side inverter maintains DC-link capacitor voltage constant and controls the reactive power delivered to the grid. The DC link created by the capacitor in the middle is required to sustain stabilized generator output to connect the power grid through an inverter circuit. It decouples the operation of both the converters, thus allowing their design and operation suits for optimization. The full-scale back-to-back converter makes constant voltage possible and fixed frequency on grid side though the rotor runs at varying speeds. The two back-to-back converters are controlled independently through decoupled d-q vector control approach.

A three-phase permanent magnet synchronous machine with sinusoidal back EMF of rated capacity 12 kW, 560 V, and 1700 rpm has been taken for analysis. The average wind speed is set as 9 m/s and subject to change between 5 and 12 m/s; it is done by either setting the limits in the saturation block or setting the limits of step input. The outputs of this block are mechanical torque T_m, mechanical power P_m, power coefficient C_p, and the tip speed ratio Z. The mechanical torque is multiplied with gain to get an electrical torque. The electromagnetic torque developed is the main source of input to rotate the PMS generator and hence to measure phase currents in the rotor terminals. It includes subsystems of PMSG with its rectifier circuit, a three-phase uncontrolled diode full-bridge type. The performance of PMSG with its steady-state and transient-state parameters has been analyzed for both open-loop and closed-loop control modes in the following sections.

As PMSG is a variable speed generator and coupled with the wind turbine without gearbox, it is not that much difficult to control the rotor speed in open-loop control method. The generator model is represented in synchronously rotating d-q reference frame. The back-to-back voltage source converters (VSC) are controlled independently through decoupled d-q vector control method. The generator-side converter regulates the speed of PMSG to implement MPPT control, that is, the electromagnetic torque of PMSG is controlled with respect to generator speed such as to achieve maximum power point. The speed control is realized through field orientation where the q-axis current is used to control the rotational speed of the generator with respect to the varying wind speed. In order to obtain maximum torque per ampere and to minimize the resistive losses in generator, the d-axis current is set to zero, while the q-axis current reference is determined by the power controller. A random source with multiplier is used to adjust wind speed to get 12–15 m/s. The model of PMSG in open-loop control mode with rectifier and two boost converters along with the inverter has been shown in **Figure 2**.

The magnitude of output voltage at stator terminals increases and the generated stator voltage is used to meet the load demand. The DC-link voltage in the intermediate stages is boosted up and measured. Whenever there is a low wind speed, back-to-back power converter draws more power from the grid to drive the generator such as to provide high startup torque. This converter decouples the wind turbine and grid, regulates the operational speed of wind turbine generator, and controls the active and reactive powers injected into the grid, thus improving the power quality.

As the voltage and frequency of generator output change along with the variations of wind speed change, the generator-side boost converter is used to track the maximum wind power. To maintain a constant switching frequency within the converter, the d- and q-axis currents are controlled indirectly through a current-regulated voltage source PWM converter. The d-q voltage control signals of the converter are obtained by comparing the d- and q-axis reference currents with the actual d- and q-axis currents of the stator. Final control action is done with d- and q-axis voltage control signals. Thus, generator/machine-side converter controls PMSG to achieve optimum energy extraction from the wind.

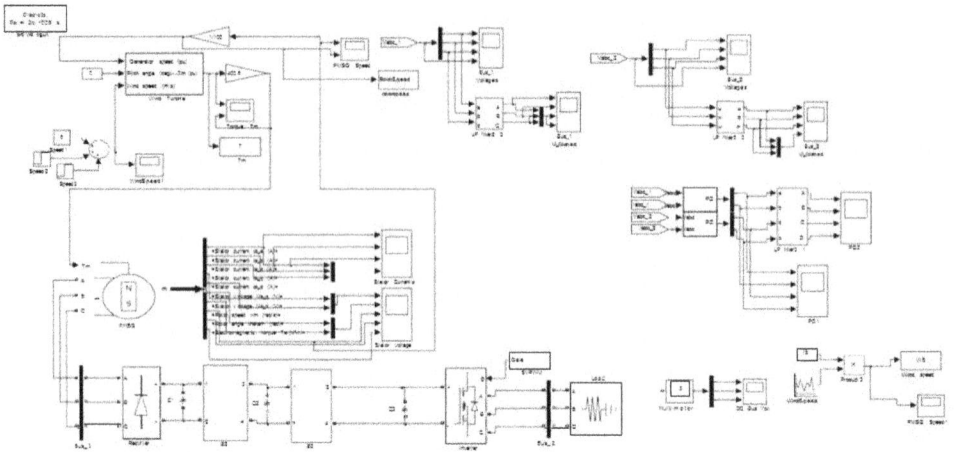

Figure 2. Model of PMSG-based WECS in open-loop control.

3. Open-loop control of PMSG-based WECS with SVM

In order to achieve variable speed operation in PMSG with the maximum power efficiency, the inverter voltage should be regulated. SVM-based grid-side inverter has been implemented to maintain the inverter voltage constant irrespective of the wind speed variations. Space-vector modulation is used to enhance the inverter voltage by selecting a revolving voltage reference vector. Eight voltage vectors in a complex $\alpha\beta$ plane, among them six active nonzero vectors (V_1–V_6) and two zero vectors (V_0 and V_7), form the look up table such as to vary the switching time of the inverter.

The main idea of this control is to transfer all the active power generated by the wind turbine to the grid and to produce no reactive power such that unity power factor is obtained. The expression for active power in d-q reference frame is given in Eq. (3) as

$$P_{dq} = \frac{3}{2}\left(v_{ds}i_{ds} + v_{qs}i_{qs}\right) \tag{3}$$

The active power is the power which is transformed to electromechanical power by the machine and expressed in Eq. (4) as

$$P_{em} = \frac{3}{2}\left(e_d i_{ds} + e_q i_{qs}\right) \tag{4}$$

$$e_d = -\omega_e L_q i_{qs} - \omega_e \psi_{qs} \tag{5}$$

$$e_q = \omega_e L_d i_{ds} + \omega_e \psi = \omega_e \psi_{ds} \tag{6}$$

Also, the active power is found by Eq. (7) given as

$$P_{em} = \frac{3}{2}\omega_e\left(\psi_d i_{qs} - \psi_q i_{ds}\right) \tag{7}$$

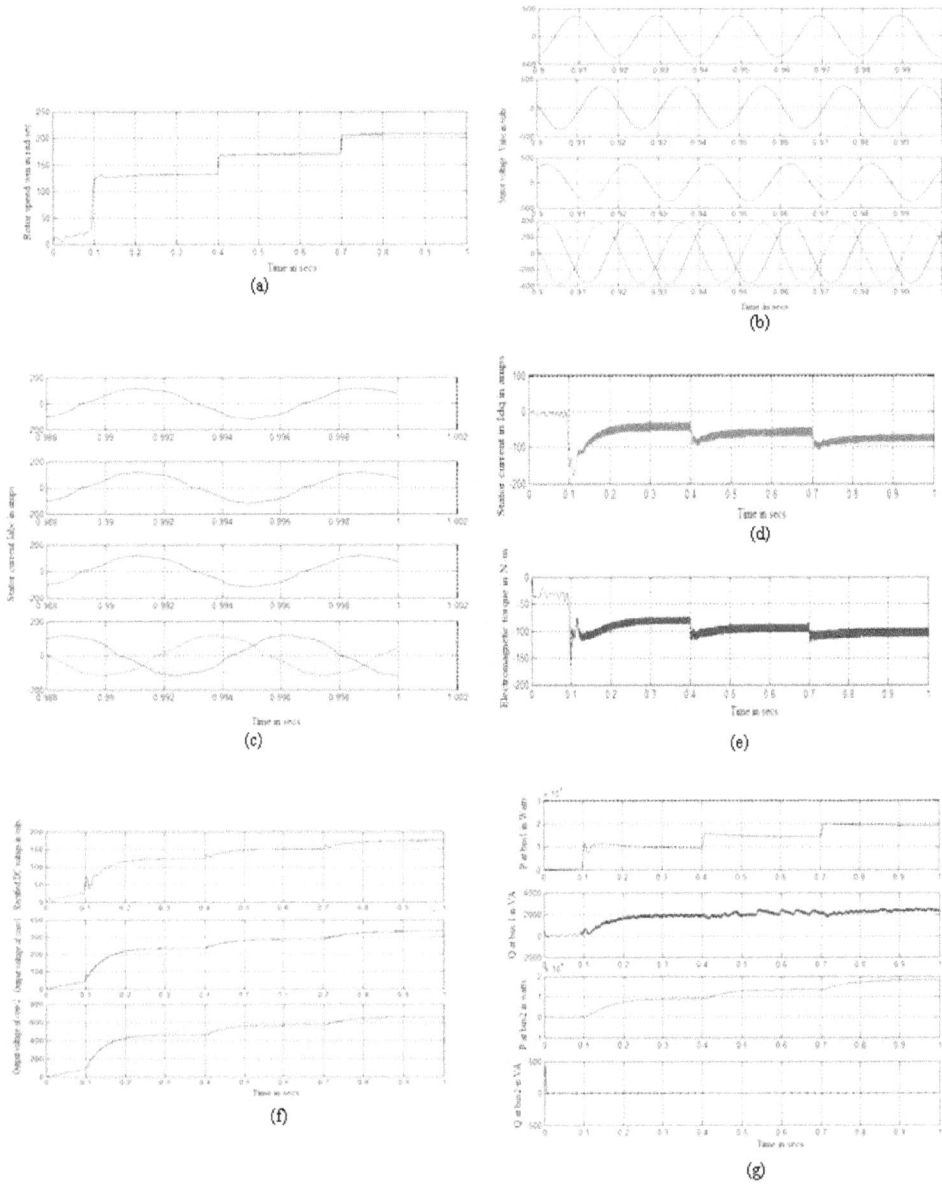

Figure 3. Simulated outputs for open loop control of PMSG based WECS. (a) Rotor speed ω_m in rad/s, (b) stator voltage V_{abc} in volts, (c) stator current I_{abc} in amps, (d) stator current I_{dq} in amps, (e) electromagnetic torque in N-m, (f) DC bus voltage at various stages in volts, and (g) real and reactive powers in stator terminals and load bus.

The relationship between ω_r and ω_m is expressed as in Eq. (8)

$$\omega_r = \frac{p}{2}\omega_m \qquad\qquad (8)$$

where

v_{ds}, v_{qs}: voltages in d-q axis reference frame w.r.t stator

i_{ds}, i_{qs}: currents in d-q axis reference frame w.r.t stator

ϕ_{ds}, ϕ_{qs}: flux linkages in d-q axis reference frame w.r.t stator

ω_e: electrical angular speed of stator flux in rad/s

ω_r: electrical speed of rotor in rad/s

ω_m: mechanical speed of rotor in rad/s

P: number of poles in the machine

Further, the generator torque is controlled by quadrature current component directly. Active and reactive power control is achieved by controlling direct and quadrature current components at the stator terminals, respectively. The operation and control of grid-side inverter is quite similar to that of controlling the rotor-side converter at the generator end. Two control loops are used to control the active and reactive power, respectively. An outer DC voltage control loop is used to set the d-axis current as reference for active power control. This assures that all the power coming from the rectifier is instantaneously transferred to the grid by an inverter.

The simulated results of PMSG-based WECS for the open-loop control mode have been depicted in **Figure 3** from (**a–f**).

From the simulated outputs, it is observed that the electromagnetic torque dips to a negative value of 100 N-m with the rotor speed at 200 rad/s. The negative sign implies that the machine was operated as a generator. The stator current gradually builds up, and, after 0.9 s, it settles down, while the stator d-q axis current peaks and gradually attains its steady state value of −100 A when time elapses.

The electromagnetic torque rises and maintains at a constant value. The idea of inserting boost converter stages in between the DC link is to effectively increase the generated AC output voltage. The generated voltage gets rectified, doubled up in first converter and raises triple-fold in the second converter and reaches to 600 V. In **Figure 3g**, it is clear that the magnitude of real and reactive power at the stator terminals and load bus almost reaches their nominal value within the stipulated time interval of 1 s.

4. Closed-loop control of PMSG-based WECS with fuzzy logic controller

Since there is no rotor coil to provide mechanical damping during transient conditions, the operational behavior of PMSG is poor in open-loop scalar V/Hz control. In order to improve the operational characteristics, to obtain a faster response, to optimize power to a greater extent, and to mobilize load power management, closed-loop control of PMSG has been attempted. Though often conventional PI controllers are preferred, due to their simple operation, easy design, and effectiveness towards linear systems, it generally does not suit for nonlinear systems of higher order, time-delayed, particularly complex, and vague systems that have no precise mathematical models.

To overcome these difficulties, fuzzy logic controllers are introduced along the intermediate stages which lie between the generator and grid.

The proposed system includes wind turbine with PMSG constituting a diode rectifier, two boost converters, two fuzzy logic controllers, an inverter, a battery, and a dump resistor. The generated AC voltage passes through the rectifier and gets converted into corresponding DC voltage. Two boost converters are used to boost the rectified output voltage obtained from PMSG and passed through the inverter through DC link. The block diagram of closed-loop control of PMSG with fuzzy logic controller is shown in **Figure 4**. The DC voltage is converted into three-phase AC voltage using a voltage source inverter comprising more number of metal-oxide-semiconductor field-effect transistors (MOSFETs) and is connected to RL load. Fuzzy logic controllers are used particularly to track the maximum power point and to promote power management. If fine and favorable wind condition prevails, and the wind speed is within the cutoff region, this autonomous wind system can meet the required load demand. On the other hand, if there is excess wind power meeting the load demand, the surplus power generated would be stored either in a battery or dissipated through dump resistor according to the battery condition.

Two fuzzy logic controllers (FLCs) are used to control the duty cycle ratio of the boost converters located near the stator side of PMSG. The first fuzzy controller is used to get MPPT by varying the duty cycle of the first boost converter, thereby increasing stator voltage of the generator.

By varying the duty cycle of boost converters, the rotor speed of PMSG is controlled to achieve optimum power. To manage energy production, another fuzzy logic controller is included such as to vary the duty cycle ratio of second boost converter and thus to regulate the DC output voltage and decide the moment to either charge/discharge the battery or dissipate

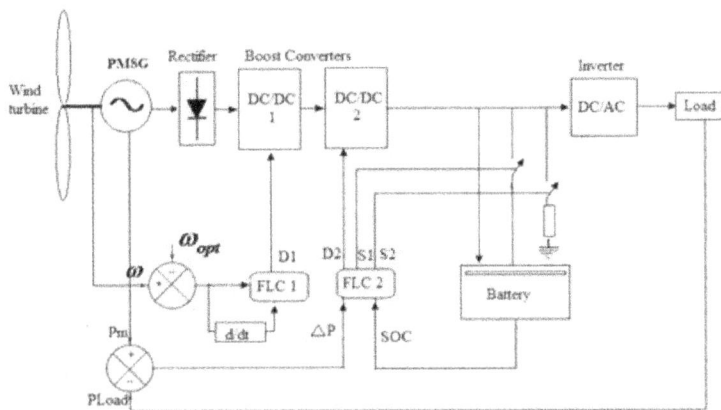

Figure 4. Block diagram of closed-loop control of PMSG-based WECS with FLC.

excess energy to the dumb resistor. With this proposed controller, wind energy is primarily provided directly to load without going through a passive element (battery). As a result, the number of charge/discharge cycles is greatly reduced, thereby extending the battery life. The Simulink model of this closed-loop control of PMSG with fuzzy logic is shown in **Figure 5**.

The first fuzzy logic controller tracks rotor speed with respect to reference speed to extract the maximum power, that is, in search of suitable generator speed which results in maximum power output. Error in speed (e) and the derivative of speed error are given as inputs and the duty cycle of the first boost converter as output is fed to the FIS editor of the first fuzzy logic controller. **Figure 6** shows the sketch of FIS editor used in the first FLC which is of Mamdani type. The duty cycle of the boost converters is changed such that the Ton and Toff periods may either increase or decrease.

Figure 6a shows the sketch of FIS editor used in the first fuzzy logic controller. The fuzzy rules are framed using "If…Then" statements with "and" operator. The membership functions of input and output variables used in the FIS of the first FLC are represented in **Figure 6b–d**.

The reference speed of the generator rotor has been taken as 1700 rpm. Fuzzy rules are formulated in a way that when the actual speed and reference speed of the rotor are almost the same, there is no need of changing the pulse width of the boost converter. The rule matrix for the first FIS editor is given in **Table 1**.

This type of FLC is mostly used in closed-loop control system, as it reduces steady-state error to zero to a greater extent. A set of 21 rules has been formulated considering the linguistic terms for the input speed error as NB, N, NS, Z, PS, P, and PB and the derivative of speed error as Negative, Zero, and Positive. Likewise, the linguistic terms used for the output of duty cycle

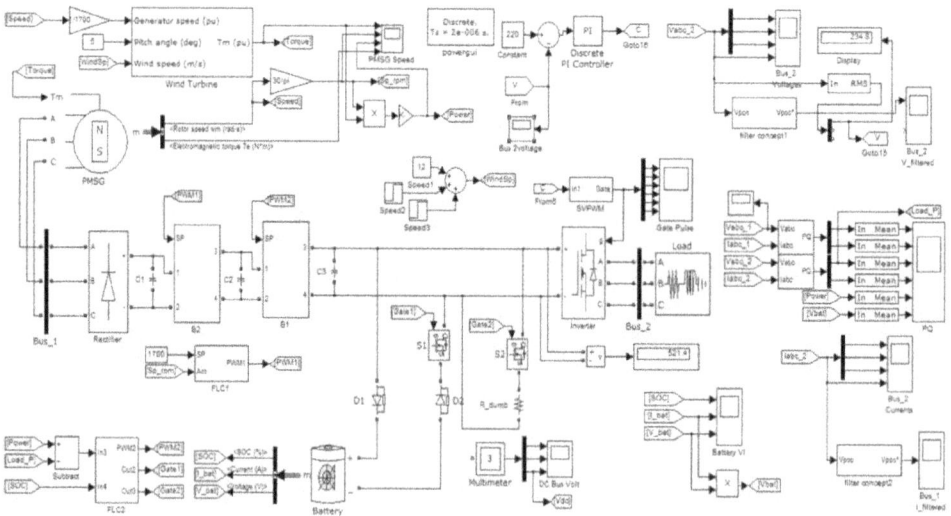

Figure 5. Model of PMSG-based WECS in closed-loop control mode with FLC.

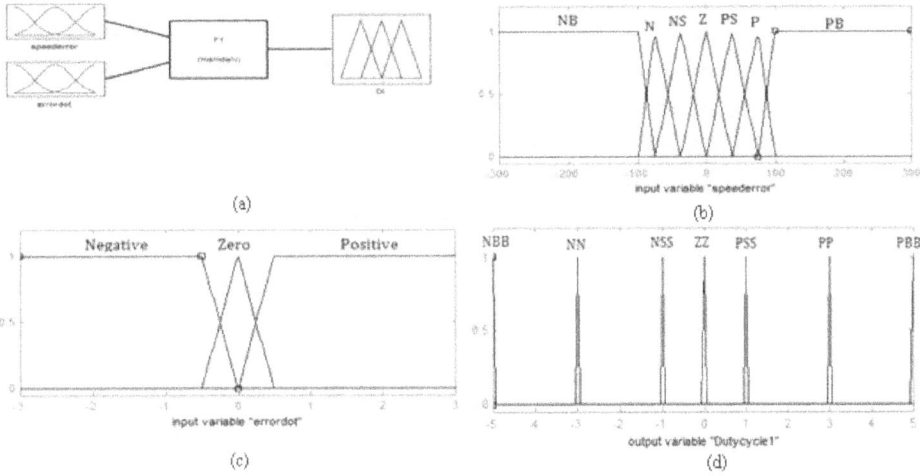

Figure 6. Membership functions used in FIS of the first FLC. (a) Mamdani-based FIS editor of the first FLC, (b) input variable of speed error, (c) input variable of speed error derivative, and (d) output variable, duty cycle 1.

Speed error/Speed error dot	Negative	Zero	Positive
NB	PBB	PP	PSS
N	PBB	PP	PSS
Z	PSS	ZZ	NSS
PS	PSS	NSS	NSS
P	NN	NSS	NBB
PB	NSS	NBB	NBB

Table 1. Rule matrix for first FLC.

of the first boost converter are PBB, PP, PSS, ZZ, NSS, NN, and NSS. The defuzzified outputs are viewed through rule viewers from the respective FIS editors.

It is explained in a manner that if the actual rotor speed is 700 rpm, the speed error becomes "Negative Big" (NB) and say, the error derivative is "Negative," then the duty cycle of the first boost converter must be "Positive Big Big" (PBB). If the actual rotor speed becomes 1000 rpm, the speed error becomes Negative (N), and the error derivative is "Negative," then the duty cycle of the first boost converter must be "PBB."

If the actual speed reaches 1500 rpm, the speed error is Negative Small (NS), the error derivative is "Negative," the duty cycle of the first boost converter must be "Positive" (P). If the actual speed exceeds the reference speed, and hence the speed error becomes "Positive Small" and the error derivative is "Negative," then the duty cycle must be "Positive Small Small" (PSS).

The second fuzzy logic controller is exclusively used for regulating the DC-link voltage by properly changing the duty cycle ratio of inverters and changing ON and OFF periods of power switches, and thus raising the power salvage. As in **Figure 7a**, the two inputs used in the second FIS editor are change in power and State of charge (SoC) of the battery. Three outputs considered are duty cycle of the second boost converter, position of switch S_1, and the position of switch S_2. The position of switch S_1 decides whether to connect or disconnect the battery to the system according to the status of the battery and the amount of excess power generated.

Position of switch S_2 determines the addition of dump resistor in the main circuit. The PWM output from each of the fuzzy logic controller serve as gating pulses for these boost converters and then to the switches too. The rules were formulated for the FIS editor of second fuzzy logic controller in such a way that if there is a change between generated power and the load power, along with the change in the SoC of the battery, duty cycle of the boost converter, position of switch S_1 and the position of switch S_2 are to be changed.

For example, If the change in power is "Negative" and SoC of the battery is "empty," then the duty cycle of the second boost converter is "VB" (Ton period is large and Toff is small), the

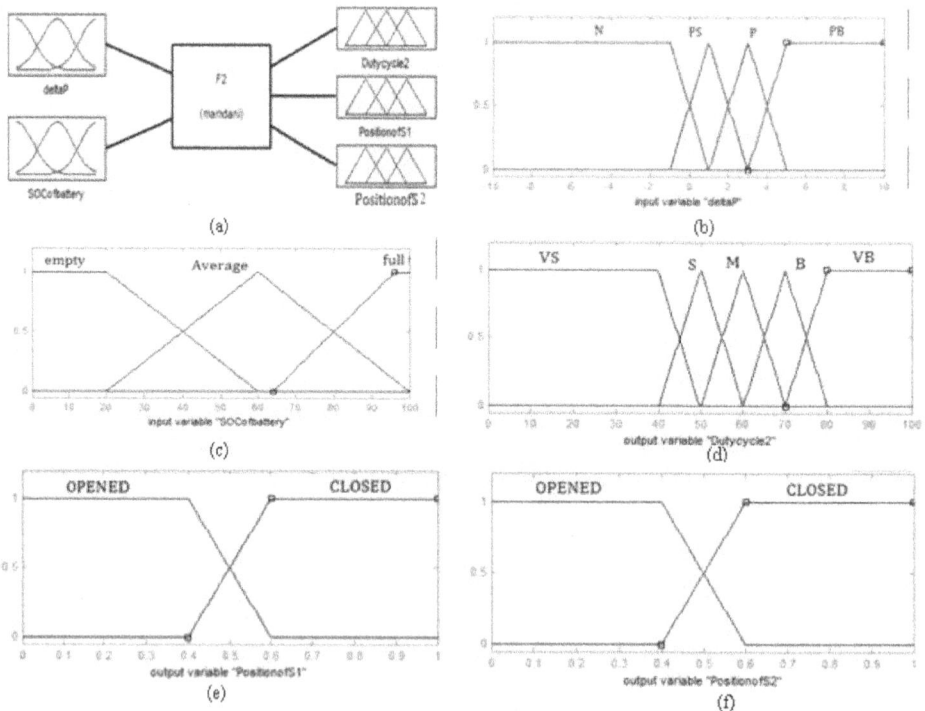

Figure 7. Membership functions used in FIS of the second FLC. (a) Mamdani-based FIS editor of the second FLC, (b) input variable delta P, (c) input variable, SoC of battery, (d) output variable, duty cycle 2, (e) output variable, position of switch S_1, (f) output variable, position of switch S_2.

position of the switch S_1 should be "opened," and the position of switch S_2 should also be "opened." If the change in power is "Negative" and the SoC of the battery is "average," then also, the duty cycle of the second boost converter is "VB," the position of the switch S_1 should be "opened," and switch S_2 should be "opened." If the change in power is "Negative" and the SoC of the battery is "full," then the duty cycle of the converter is "VB," the position of switch S_1 should be "opened" and position of switch S_2 is "opened."

If not, for other conditions, the change in power is either "Positive," or "Positive small," or "Positive big," the duty cycle of the second converter, position of switch S_1, and position of switch S_2 as "opened" or "closed" and varied correspondingly. The linguistic terms used as inputs and outputs are stated as follows:

For the inputs of change in power as N, PS, P, and PS, and battery's state-of-charge (SoC) as empty, average, and full, the corresponding outputs, namely duty cycle of second boost converter as VS, S, M,B, and VB, position of switch S_1 and switch S_2 as opened and closed with their membership functions are shown in **Figure 7**.

By framing 36 fuzzy "if-then" rules, the duty cycle of the second boost converter is changed and position of switches being altered to prevent wastage and dissipation of power in loads. For analysis, it has been taken such that the load power to be met is around 6 kW and the wind power fluctuates between 5.1 and 6.5 kW. Initially, it is assumed that the wind power is 5.1 kW. If the change in power is "negative" and the SoC of the battery is "empty," then the duty cycle of the second converter should be "Very Big" (VB). If the change in power is "negative" and SoC of battery is "average," then the duty cycle of the second converter should be "VB."

SOC of battery/ Change in power	Empty	Average	Full
N	VB	VB	VB
	Open	Open	Open
	Open	Open	Open
PS	B	S	VS
	Close	Close	Open
	Open	Open	Close
P	M	S	VS
	Open	Open	Open
	Open	Open	Open
PB	S	VS	VS
	Close	Close	Open
	Open	Open	Close

Table 2. Rule matrix for the second FLC.

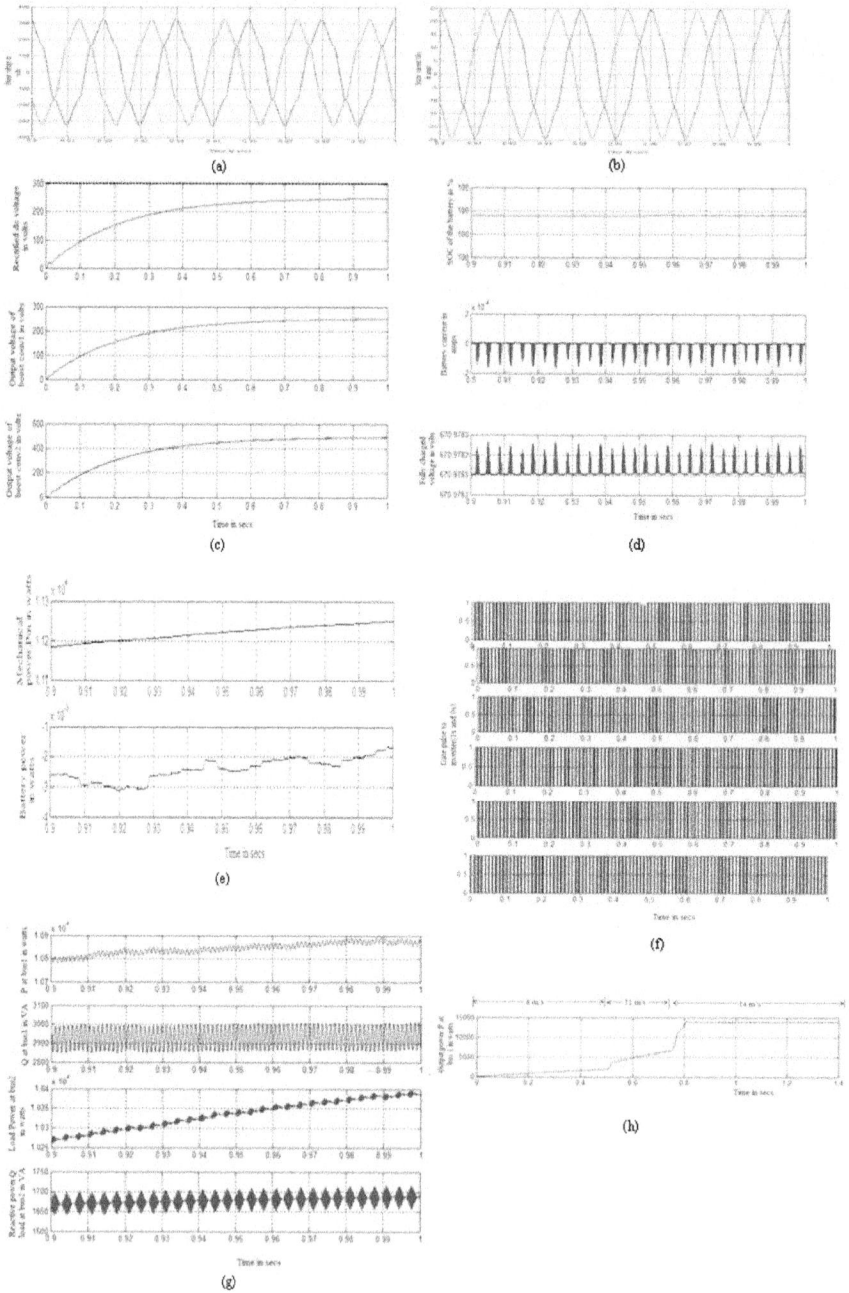

Figure 8. Simulated outputs for closed-loop control of PMSG-based WECS. (a) stator voltage V_{abc} in volts, (b) stator current I_{abc} in amps, (c) DC-link voltage in the intermediate stages of boost converter, (d) battery parameters, (e) mechanical power and battery power, (f) gate pulses to inverter (t_{on} and t_{off}), (g) generated power and load power at bus 1 and bus 2 and (h) generated output power w.r.t wind speed variations.

If the change in power is "negative" and the SoC of the battery is "full," then the duty cycle of the second converter should be "VB". Conversely, if the generated wind power is very less compared to load power, it is taken that the change in power is "Negative" (5.1–6 kW).

If the obtained wind power exceeds load power, that is, 6.1 kW, then it is understood that the change in power is "Positive Small" (PS) (6.1–6 kW). Similarly, if the wind power and the load power are equal, then it is taken that the change in power is "Positive" (6.0–6 kW) and, accordingly, the other such rules are formulated. Again if the wind power is 6.5 kW, then it is implicit that the change in power is "Positive Big" (6.5–6 kW), and hence the duty cycle is "small," switch S_1 is closed to connect the system with battery and switch S_2 is opened (**Table 2**).

The steady- and transient-state characteristics of the closed-loop control mode of PMSG with fuzzy logic controller through which the most important parameters derived are depicted in **Figure 8** from (**a–f**).

5. Conclusion

From the simulated results, it is observed that the load requirements are directly met by the generated power itself but not by the battery power when the wind power is plenty. The mechanical power reaches its nominal value of 11.3 kW. The simulated results for both open-loop and closed-control modes have been compared and tabulated in **Table 3**. The generated power P at the load bus terminals for closed-loop control mode is greater than the power attained in the open-loop control mode.

Also, it seems that there is a steady increase in the output power, while it is fluctuating in the open-loop control mode with respect to the wind speed. According to the variations in the wind velocity, there is a remarkable change observed in the output power generated in bus 1. When the wind speed is of 8 m/s, the power generated attains 5 kW while it fluctuates between 9 and 11 m/s, there is a gradual increase in power. It reaches nearly the peak value of 14 kW, which is undesirable as wind speed goes beyond 12 m/s.

When the wind speed fluctuates between 8 and 12 m/s, the reactive power is zero initially and reaches a constant value of 2200 VA in its steady state. The speed curve of the rotor attains its steady-state value when the time t is 0.5 s. Also, it is observed that the terminal voltage in the stator in all three phases is regulated and boosted up to 330 V in the first stage and then to 635 V in the second stage, and the three-phase stator current is found to be around 25 A.

During the closed-loop control mode, though drastic change in the power has not been observed in the boost converter stages, the generated power exceeds the load power, and hence apart from meeting the load demands, it supplies the battery load which is clearly visible from its charging current. Also, the battery voltage is getting its full rated value of 670 V. Further, the generated stator voltage is boosted up, and hence optimum power is observed in both grid and loads, that is, the generated voltage at the stator terminals gradually builds up and attains its steady-state value after 0.9 s. The magnitude of generated power at the load bus

Steady state electrical parameters/control technique used	Open loop control mode	Closed loop control mode with fuzzy logic
Stator current, I_{abc}	100 amps	24.4 amps
dq axes current in stator, I_{dq}	-100.4 amps	-40 amps
Stator current, I_{abc} after boost converter stage	--	77 amps
Stator voltage V_{abc} after boost converter stage	320 volts	348 volts
Mechanical Torque, T_m	-60 N-m	-49 N-m
Electromagnetic Torque T_{em}	-100 N-m	-50 N-m
Rotor Speed, w_m	210.5 rad/sec	74 rad/sec
DC bus voltage V_{dc} (rectified)	174 volts	240 volts
DC bus voltage V_{dc} (boost converter stage1)	330 volts	250 volts
DC bus voltage V_{dc} (boost converter stage2)	635 volts	490 volts
Battery SOC	--	100
Battery current	--	-256 amps
Battery voltage	--	670.97 volts
Battery power	--	Almost zero
Generated power P@bus1	1×10^4 watts	1.05×10^4 watts
Reactive power Q @bus1	2200 VA	2900 VA
Load power, P@bus2	--	1.04×10^4 watts
Reactive power, Q@bus2	--	1725 VA

Table 3. Simulated results for PMSG-based WECS.

seems increasing around 10.4 kW while reactive power reaches 1725 VA but with some ripples. The mechanical power also gets its steady value of 11.2 kW within a time frame of 1 s.

In addition, the load power management is excellent in closed-loop control method using fuzzy logic controller. Also, if the generated power is found in excess, it would be stored in the battery for future power salvage. There is a notable increase in the magnitude of stator voltage and hence the generated power; ripples are considerably reduced compared to open-loop control mode, thus improving the power quality as well.

Author details

Anbarasi Jebaselvi Jeya Gnanaiah David* and Meenakshi Veerappan

*Address all correspondence to: anbarasi.jebaselvi@gmail.com

Faculty of Electrical and Electronics Engineering, Sathyabama Institute of Science and Technology (Deemed to be University), Chennai, Tamil Nadu, India

References

[1] Errami Y et al. Modeling and control strategy of PMSG based variable speed wind energy conversion system. In: International Conference on Multimedia Computing and Systems (ICMCS), Morocco: COS ONE Ouarzazate. 7-9th Apr 2011. pp. 1-6

[2] Wang Y, Hai R. Power control of permanent magnet synchronous generator directly driven by wind turbine. International Journal of Signal Processing Systems. Dec 2013; 1(2):244-249. DOI: 10.12720/ijsps.1.2.244-249

[3] Melicio R, Mendes VMF, Catalão JPS. A Pitch Control Malfunction Analysis for Wind Turbines with PMSG and Full-power Converters: Proportional Integral Versus Fractional Order Controllers. Covilha: University of Beira Interior; 2009. pp. 465-494

[4] Meenakshi V, Paramasivam S. Modeling and simulation of PMSG using SVPWM switching technique. International Journal of Applied Engineering Research. 2015;10(6): 5165-5171

[5] Yenduri K, Sensarma P. Maximum power point tracking of variable speed wind turbines with flexible shaft. IEEE Sustainable Energy. July. 2016;7(3):956-965

[6] Morimoto S et al. Sensorless output maximization control for variable-speed wind generation system using PMSG. IEEE Transactions on Industrial Applications. 2005;41(1):60-67

[7] Kazmi SMR et al. A novel algorithm for fast and efficient speed-sensorless maximum power point tracking in wind energy conversion systems. IEEE Transactions on Industrial Electronics. 2011;58(1):29-36

[8] Haque ME et al. A novel control strategy for a variable-speed wind turbine with a permanent magnet synchronous generator. IEEE Transactions on Industrial Applications. 2010;46(1):331-339

[9] Tan K, Islam S. Optimum control strategies in energy conversion of PMSG wind turbine system without mechanical sensors. IEEE Transactions on Energy Conversion. 2004; 19(2):392-399

[10] Hemeida AM, Farag WA, Mahgoub OA. Modeling and control of direct driven PMSG for ultra large wind turbines. World Academy of Science, Engineering and Technology. 2011;5:11-27

[11] Rolan A, Luna A, Vazquez G, Aguilar D, Azevedo G. Modeling of a variable speed wind turbine with a permanent magnet synchronous generator. In: IEEE International Symposium on Industrial Electronics (ISIE-2009). Seoul, Korea. 2009. pp. 734–739

[12] Mahersi E, Khedher A, Faouzi Mimouni M. The wind energy conversion system using PMSG controlled by vector control and SMC strategies. International Journal of Renewable Energy Research. 2013;3(1):41-50

[13] Huang N. Simulation of power control of a wind turbine permanent magnet synchronous generator system [Master's thesis report]. Milwaukee: Marquette University; 2013

[14] Tan K, Islam S. Optimum control strategies in energy conversion of PMSG wind turbine system without mechanical sensors. IEEE Transactions on Energy Conversion. 2004;**19**(2): 392-399

[15] Melicio R, Mendes VMF, Catalao JPS. A pitch control malfunction analysis for wind turbines with PMSG and full-power converters: Proportional integral versus fractional-order controllers. Electric Power Components and Systems. Taylor and Francis Ltd. Jan 2010;**38**(4):387-406

[16] Melicio R, Mendes VMF, Catalao JPS. Wind turbines with permanent magnet synchronous generator and full-power converters: Modeling, control and simulation. Wind Turbines, InTech. 2011. pp. 465-495. ISBN: 978–953-307-221-0. Available from: http://www.intechopen.com

[17] Yin M, Li G, Zhou M, Zhao C. Modeling of the wind turbine with a permanent magnet synchronous generator for integration. IEEE, Power Engineering Society General Meeting. Tampa. 24-28 June 2007:1-6

[18] Chinchilla M, Arnaltes S, Burgos JC. Control of permanent-magnet generators applied to variable-speed wind-energy systems connected to the grid. IEEE Transactions on Energy Conversion. 2006;**21**(1):130-135

[19] Stroe DI, Stany AI, Visa I, Stroe I. Modeling and control of variable speed wind turbine equipped with PMSG. In: 13th World Congress in Mechanism and Machine Science; Guanajuato, Mexico; 2011. pp. 1-5

[20] Brahmi J, Krichen L, Ouali A. A comparative study between three sensorless control strategies for PMSG in wind energy conversion system. In: Advanced Control and Energy Management Research Unit ENIS. Sfax: Department of Electrical Engineering, University of Sfax. 2009

[21] Samanvorakij S, Kumkratug P. Modeling and simulation PMSG based on wind energy conversion system in MATLAB/SIMULINK. In: Proceedings of the Second International Conference on Advances in Electronics and Electrical Engineering. United States: AEEE; 2013

[22] Shariatpanah R et al. A new model for PMSG based wind turbine with yaw control. IEEE Transactions on Energy Conversion. Dec 2013;**28**(4):929-937

[23] Chen Y et al. A control strategy of direct driven permanent magnet synchronous generator for maximum power point tracking in wind turbine application. In: International Conference on Electrical Machines and Systems. Hankou Wuhan, China. 2008

[24] Ahmed W, Ali SMU. Comparative study of SVPWM (space vector pulse width modulation) and SPWM (sinusoidal pulse width modulation) based three phase voltage source inverters for variable speed drive. In: ICSICCST 2013 IOP Conference Series: Materials Science and Engineering. Vol. 51. Karachi, Pakistan. 24-26th June 2013

[25] Sharma S, Shrivastav A, Rawat H, Warudkar V. Modeling and control of permanent magnet synchronous generator connected to grid driven by wind turbine using fast

Simulator and SVPWM Technique. IOSR Journal of Electrical and Electronics Engineering (IOSR-JEEE). July–August 2014;9(4) Ver. IV:61–71 e-ISSN: 2278-1676, p-ISSN: 2320-3331, www.iosrjournals.org

[26] Rosyadi M, Muyeen SM, Takahashi R, Tamura J. A design fuzzy logic controller for a permanent magnet wind generator to enhance the dynamic stability of wind farms. Applied Sciences. 2012;**2**:780-800. DOI: 10.3390/app2040780 ISSN 2076-3417

[27] Trinh QN, Lee HH. Fuzzy logic controller for maximum power tracking in PMSG based wind power systems. In: Huang DS et al., editors. ICIC 2010, LNAI 6216, 18-21st Aug 2010 Hunan Province, China Changsha, China. Berlin: Springer-Verlag; 2010. pp. 543-553

[28] Vanitha K, Shravan C. Permanent magnet synchronous generator with fuzzy logic controller for wind energy conversion system. International Journal of Engineering Research & Technology (IJERT). November 2013;**2**(11):3943-3950

[29] Jane Justin B, Rama Reddy S. Fuzzy controlled SEPIC based micro wind energy conversion system with reduced ripple and improved dynamic response. Journal of Electrical Engineering, pp. 1-11. ISSN: 1582-4594. http://www.jee.ro/

[30] Harrabi N, Souissi M, Aitouche A, Chaabane M. Intelligent control of wind conversion system based on PMSG using T-S fuzzy scheme. International Journal of Renewable Energy Research. 2015;**5**(4):952-960

High Voltage Transmission Line Vibration: Using MATLAB to Implement the Finite Element Model of a Wind-Induced Power-Line Conductor Vibration

Chiemela Onunka and Evans Eshiemogie Ojo

Additional information is available at the end of the chapter

http://dx.doi.org/10.5772/intechopen.75186

Abstract

Wind-induced vibration affects the performance and structural integrity of high voltage transmission lines. The finite element method (FEM) is employed to investigate wind-induced vibration in MATLAB. First, the FEM model was used to develop the equation of motion of the power line conductor. In addition, dampers, conditions for damping, free and forced vibrations of the overhead conductor were considered in the FEM model. Wind-induced experiments were conducted in the laboratory using an actual overhead power conductor. The developed FEM models were simulated in the MATLAB computing environment. The results from the MATLAB simulation, finite element and experimental recordings were compared in order to evaluate the efficacy of models simulated in MATLAB and developed using the FEM.

Keywords: aeolian vibration, power conductor damping, resonant frequency, MATLAB

1. Introduction

The availability and use of electrical power in the society is crucial in the development and growth of the society. Power generated from power stations is transmitted using high voltage transmission lines. The transmission of power from the point of generation to the point of use requires complex network of high voltage lines, systems and components [1]. High voltage conductors are usually subjected to vibration and the vulnerabilities of the power lines to vibration can lead to fatigue failure. Thus, power loading determination and control on the power grid can influence the integrity of the transmission network. High voltage conductor vibration is very difficult to model due to the fact that the responses exhibit a non-linear behaviour. There has been concerted effort to try and predict the conductor response as a result of aeolian vibration. Evaluations of conductor vibration caused by aeolian forces have been investigated be several researchers [2–6].

Recent researches developed models to investigate wind-induced vibration using nonlinear time history, expert systems, the concept of principal modes, aero-elastic and bending stiffness [7–17]. The investigation of vortex formation and the phenomenon of wind-induced vibration was done using the concept fluid–solid dynamic excitation [18–20]. The models were used to determine how wind loading influenced the oscillation of transmission lines. This form of investigation was done by experimental studies carried out in a wind tunnel [21, 22]. The outcomes of these experiments were used to determine conductor input loading. Several models developed by various researchers can be used to determine conductor damping and also the placement of vibration absorbers on the line conductors to curtail the effect of cable mechanical oscillation [23–27]. Based on the various models that have been developed by researchers as indicated in the first and second paragraphs, there is a need to further analyze wind-induced vibration using finite element method (FEM) in MATLAB.

The design, construction and maintenance of power transmission network requires adequate understanding of the system dynamics that occurs when subjected to vortex induced vibration [28]. Various analysis can be conducted using techniques that suit certain objectives. System integrity in high voltage transmission lines is of paramount importance. MATLAB is a multi-model simulation environment used for numerical computing. It can be integrated with physical hardware or systems in order to determine real-time performance, characteristics and behavior. MATLAB also provides a platform for special hardware in loop simulations [29]. These functionalities amongst others are vital in determining various characteristics and behaviors in high voltage transmission lines.

High voltage transmission lines and grid can experience vulnerabilities such as vibration, electromagnetic transients, fatigue, transmission loss, switching surges, conductor sag fluctuation [30, 31]. When the conductor experiences vibration, the transmission lines experience high amplitudes of vibrations from wind forces and can lead to fatigue of the transmission lines [1]. The use of systems simulation and analysis provides the platform to understand the response of the transmission conductor. The responses considered in the chapter include transmission line excitation through wind loading, conductor properties such as damping and damper placement used in mitigating the vibration.

The chapter discussed the development and implementation of a wind-induced high voltage transmission line vibration using finite element method (FEM) in MATLAB. The sections in the chapter discussed the development of transmission line equation of motion, the solution to the equation of motion, free and forced vibration of the transmission line, dampers and conductor self-damping, FEM MATLAB setup and implementation, simulation of FEM models. The chapter also discussed results from FEM models, simulation and experimental investigation. The chapter is focused towards the development of a finite element method and its implementation on the MATLAB software. The developed finite element method (FEM) was based on the concept of the simply supported beam model and it was used in modeling the transverse vibration of power line conductors. The results from the FEM were then compared with results from the analytical model and results obtained from experimental studies documented in [1].

The results from MATLAB simulations from the finite element models and experimental results were compared in order to determine the accuracy of the models. The developed FEM was then used as the means to verify the effect of varying the conductor axial tension on the natural frequencies of the conductors.

2. Transmission line equation of motion (EOM)

The transverse displacement of high voltage transmission line conductor is generally caused by wind loading. This form of vibration with small displacement is known as aeolian vibration and it is a source of concern to the power lines reliability. One the vulnerabilities is that it can cause fatigue failure of the transmission lines. Conductors are example of continuous or distributed systems and modeling its mechanical vibration can either be as a beam or taut string. In [18, 19], it was ascertained that modeling a conductor as a beam is more accurate than modeling it as a taut string due to the effect of the bending stiffness. Hence, in line with the above, the conductor transverse vibration was modeled as a beam, simply supported or pinned at both ends. The distributed loading on the conductor is replaced by effective point load that can effectively have the same resultant effect as that of the actual distributed load.

The high voltage transmission line equation of motion was formulated by assuming that power conductors can modeled as beams with fixed ends. The following assumptions were considered [1]:

- The power conductor is uniform along its length and it is slender

- The power conductor is a solid with cylindrical body having both linear and homogeneous physical properties throughout its cross-sectional area

- The power conductor has a symmetrical plane which acts as the plane of vibration such that there is the decoupling of translational and rotational motion.

The assumptions were based on beam theory. In considering the power conductor as a beam, sagged by a tensile force S, being acted upon by a concentrated wind load $f(x, t)$, with cross-sectional area A, density ρ, flexural rigidity EI, displaced at a distance of x after time. In Eq. (1), the high voltage transmission line equation of motion is expressed as:

$$f(x, t) = EI \frac{\partial^4 y(x, t)}{\partial x^4} - S \frac{\partial^2 y(x, t)}{\partial x^2} + \rho A \frac{\partial^2 y(x, t)}{\partial t^2} \tag{1}$$

For $x \in (0, 1)$, $t > 0$. The boundary conditions are expressed and indicated in Eqs. (2) and (3):

$$y(0, t) = \frac{\partial^2 (0, t)}{\partial x^2} = 0 \tag{2}$$

$$y(l, t) = \frac{\partial^2 (l, t)}{\partial x^2} = 0 \tag{3}$$

The initial conditions at $t = 0$ are indicated in Eq. (4), Eq. (5) and expressed as:

$$y(x, 0) = y_o(x) \tag{4}$$

$$\dot{y}(x, 0) = \dot{y}(x) \tag{5}$$

Introducing the mass per unit length of the power conductor, the new equation of motion indicated in Eq. (6) is expressed as:

$$f(x, t) = EI\frac{\partial^4 y(x, t)}{\partial x^4} - S\frac{\partial^2 y(x, t)}{\partial x^2} + m\frac{\partial^2 y(x, t)}{\partial t^2} \tag{6}$$

In order to derive a possible solution, the model was simplified using dimensionless functions and Dirac delta functions. In Eqs. (7)–(12), the variables are expressed in dimensionless form and expressed as:

$$Y = \frac{y(x, t)}{D} \tag{7}$$

$$X = \frac{x}{L} \tag{8}$$

$$\tau = \frac{t}{f} \tag{9}$$

$$I_p = \frac{Df^2}{g} \tag{10}$$

$$S_p = \frac{SD}{\gamma L^2} \tag{11}$$

$$M_p \frac{EI \; D}{\gamma L^4} \tag{12}$$

Eq. (13) indicates the revised equation of motion and it is expressed as:

$$M_p \cdot \frac{\partial^4 Y}{\partial X^4} - S_p \frac{\partial^2 Y}{\partial X^2} + I_p \frac{\partial^2 Y}{\partial \tau^2} = \frac{1}{\gamma}\left[F(X, \tau) + \sum_n \delta(X - X_n)F_n(\tau) \right] \tag{13}$$

Where γ represents the power conductor weight per unit length and g represents gravitational constant. $X_n\delta(X - X_n)$ represents the Dirac delta function, $F(X, \tau)$ denotes the net transverse force per unit length acting on the conductor and $F_n(\tau)$ denotes the nth concentrated force acting transversely on the conductor.

3. Solution to the EOM

The general solution to the high voltage transmission line equation of motion was derived using Euler-Bernoulli equation. The particular solution to the equation of motion was derived using a product of two functions. The two functions were first separated using the principle of variable separation as expressed in Eq. (14) [19]:

$$Y(x,t) = X(x)T(t) \tag{14}$$

Where $X(x)$ is the normalized function representing the mode shape of the equation of motion. The normalized function ensures that orthogonality condition was satisfied in the derivation of the EOM model solution. Applying the normalized function in the EOM yields Eqs. (15) and (16):

$$EI \overset{////}{X}(x) - S \overset{//}{X}(x) - \omega^2 \rho AX(x) = 0 \tag{15}$$

$$\ddot{T}(t) + \omega^2 T(t) = 0 \tag{16}$$

Where $\overset{////}{X}(x) = \frac{d^4 y}{dx^4}$, $\overset{//}{X}(x) = \frac{d^2 y}{dx^2}$, $\ddot{T}(t) = \frac{d^2 y}{dt^2}$ and ω^2 is a constant that equates x and t. Assuming that $X(x) = Ze^{\Psi x}$, the model is expressed in Eq. (17) as:

$$Ze^{\Psi x} \left(EI\Psi^4 - S\Psi^2 - \rho A\omega^2 \right) = 0 \tag{17}$$

Considering that $Ze^{\Psi x} \neq 0$, hence $\left(EI\Psi^4 - S\Psi^2 + \rho A\omega^2 \right) = 0$. The general solution of the Euler-Bernoulli equation which represents the solution to the equation of the motion of the transmission line is expressed in Eqs. (18) and (19) as [9]:

$$\Omega^2, \Psi^2 = -\frac{(-S) \pm \sqrt{S^2 - 4(EI)(-\rho A\omega^2)}}{2EI} \tag{18}$$

$$\Omega, \Psi = (\pm)\sqrt{\frac{S \pm \sqrt{S^2 + 4EI(\rho A\omega^2)}}{2EI}} \tag{19}$$

The values of Ω and Ψ represents the general solution of the equation of motion. The practical implication of the derived solution is that it represents the transverse vibration of the high voltage transmission line. The derived solution has infinite number of solutions and the solution is indexed to accommodate all the possible solutions from the model. The indexed solution is expressed in Eqs. (20) and (21) as:

$$\Omega_n = \sqrt{\frac{S}{2EI} + \sqrt{\frac{S^2}{(2EI)^2} + m_L \frac{(2\pi f_n)^2}{EI}}} \tag{20}$$

$$\Psi_n = \sqrt{-\frac{S}{2EI} + \sqrt{\frac{S^2}{(2EI)^2} + m_L \frac{(2\pi f_n)^2}{EI}}} \tag{21}$$

Where $\omega_n = 2\pi f_n$ and for $n = 1, 2, 3, \ldots$.

In Eqs. (22)–(24), the infinite natural frequencies of the power conductor were derived while considering that the mode shape is the same as a pinned-pinned beam eigenfunction model with no external force. Hence,

$$Y_n(x,t) = \sin\frac{n\pi x}{l}\cos\omega_n t \tag{22}$$

$$EI\left(\frac{n\pi}{l}\right)^4\sin\frac{n\pi x}{l}\cos\omega_n t - S\left(-\frac{n\pi}{l}\right)^2\sin\frac{n\pi x}{l}\cos\omega_n t + \rho A(-\omega_n)\sin\frac{n\pi x}{l}\cos\omega_n t = 0 \tag{23}$$

$$\sin\frac{n\pi x}{l}\cos\omega_n t\left[\frac{EI}{\rho A}\left(\frac{n\pi}{l}\right)^4 + \frac{S}{\rho A}\left(\frac{n\pi}{l}\right)^2 - \omega_n^2\right] = 0 \tag{24}$$

The natural frequency of the power conductor in rad/s is expressed in Eqs. (25) and (26) as:

$$\omega_n^2 = \left(\frac{n\pi}{l}\right)^2\frac{S}{A\rho} + \left(\frac{n\pi}{l}\right)^4\frac{EI}{A\rho} \tag{25}$$

$$\omega_n = \sqrt{\left(\frac{n\pi}{L}\right)^2\frac{S}{m_L}\left[1 + \left(\frac{n\pi}{L}\right)^2\frac{EI}{S}\right]} \tag{26}$$

The natural frequency in Hz is expressed in Eq. (27) as:

$$F_n = \frac{1}{2\pi}\sqrt{\left(\frac{n\pi}{L}\right)^2\frac{S}{m_L}\left[1 + \left(\frac{n\pi}{L}\right)^2\frac{EI}{S}\right]} \tag{27}$$

4. Free vibration of power conductor

The self-damping model of the power conductor provided the basis to analyze free vibration experienced by the conductor. Free vibration occurs when the forcing function causing the power conductor to vibrate become zero. Hence the equation of motion is expressed in Eq. (28) as [19]:

$$EI\frac{\partial^4 y(x,t)}{\partial x^4} - S\frac{\partial^2 y(x,t)}{\partial x^2} + \beta I\frac{\partial^5 y(x,t)}{\partial x^4 \partial t} + C\frac{\partial y(x,t)}{\partial t} + \rho A\frac{\partial^2 y(x,t)}{\partial t^2} = 0 \tag{28}$$

Applying the principle of separation of variable to the equation of motion yields Eq. (29):

$$EI\ \overset{////}{X}(x)T(t) - S\ \overset{//}{X}(x)T(t) + \beta I\ \overset{///}{X}(x)\dot{T}(t) + CX(x)\dot{T}(t) + \rho A T(t)X(x) = 0 \tag{29}$$

Integrating the eigenfunction $X_n(x) = \sin\left(\frac{n\pi x}{l}\right)$ in the model yields Eqs. (30)–(33):

$$EI\left[\left(\tfrac{n\pi}{l}\right)^4 \sin\left(\tfrac{n\pi x}{l}\right)\right]T(t) - S\left[\left(\tfrac{-n\pi}{l}\right)\sin\left(\tfrac{n\pi x}{l}\right)\right]T(t)$$
$$+\beta I\left[\left(\tfrac{n\pi}{l}\right)^4 \sin\left(\tfrac{n\pi x}{l}\right)\right]\dot{T}(t) + C\left[\sin\left(\tfrac{n\pi x}{l}\right)\right]\dot{T}(t) + \rho A\left[\sin\left(\tfrac{n\pi x}{l}\right)\right]\ddot{T}(t) = 0 \tag{30}$$

$$\sin\left(\tfrac{n\pi x}{l}\right)\left[\begin{array}{l} EI\left(\tfrac{n\pi}{l}\right)^4 T(t) + S\left(\tfrac{n\pi}{l}\right)^2 \dot{T}(t) \\[4pt] \quad + \beta I\left(\tfrac{n\pi}{l}\right)^4 T(t) + C\dot{T}(t) + \rho A\ddot{T}(t) \end{array}\right] = 0 \tag{31}$$

$$\rho A\ddot{T}(t) + \left[\beta I\left(\tfrac{n\pi}{l}\right)^4 + C\right]\dot{T}(t) + \left[S\left(\tfrac{n\pi}{l}\right)^2 + EI\left(\tfrac{n\pi}{l}\right)\right]T(t) = 0 \tag{32}$$

$$\ddot{T}(t) + \left[\tfrac{\beta I}{\rho A}\left(\tfrac{n\pi}{l}\right)^4 + \tfrac{C}{\rho A}\right]\dot{T}(t) + \left[\tfrac{S}{\rho A}\left(\tfrac{n\pi}{l}\right)^2 + \tfrac{EI}{\rho A}\left(\tfrac{n\pi}{l}\right)^4\right]T(t) = 0 \tag{33}$$

Considering that the vibration model represents a multi-degree vibration system. The natural frequency of the power conductor is determined and expressed in Eqs. (34) and (35):

$$\omega_n^2 = \frac{S}{\rho A}\left(\frac{n\pi}{l}\right)^2 + \frac{EI}{\rho A}\left(\frac{n\pi}{l}\right)^4 \tag{34}$$

$$2\xi\omega_n^2 = \left[\frac{\beta I}{\rho A}\left(\frac{n\pi}{l}\right)^4 + \frac{C}{\rho A}\right] \tag{35}$$

The temporal solution to the free vibration model is expressed in Eq. (36) as:

$$T_n = A_1 e^{-\xi_n\omega_n t}\sin(\omega_d t + \phi) \tag{36}$$

The solution can also be represented in Eq. (37) and expressed as:

$$T_n = e^{-\xi_n\omega_n t}(B_1 \sin\omega_d t + B_2 \cos\omega_d t) \tag{37}$$

Where the damped frequency of the power conductor is expressed in Eq. (38) as:

$$\omega_d = \omega_n\sqrt{1 - \xi^2} \tag{38}$$

The system response is expressed in Eq. (39) as:

$$y(x,t) = \sum_{n=1}^{\infty} A_1 e^{-\xi_n\omega_n t}\sin(\omega_d t + \varphi)\sin\frac{n\pi x}{l} \tag{39}$$

The response can also be represented in Eq. (40) and expressed as:

$$y(x,t) = \sum_{n=1}^{\infty} \left[e^{-\xi_n\omega_n t}(B_1 \sin\omega_d t + B_2 \cos\omega_d)\right]\sin\frac{n\pi x}{l} \tag{40}$$

5. Forced vibration of power conductor

High voltage transmission lines are exposed to loading from the wind. The actual system representation through system simulation strategy considers a case of distributed load through the span of the conductor. In order to simplify simulations, the external force acting on the conductor is represented as a point load. In Eqs. (41)–(43), the equation of motion is solved with an excitation force in order to evaluate the actual response of high voltage transmission lines under aeolian vibration [1]. Hence,

$$EI\frac{\partial^4 y(x,t)}{\partial x^4} - S\frac{\partial^2 y(x,t)}{\partial x^2} + \beta I\frac{\partial^5 y(x,t)}{\partial x^4 \partial t} + C\frac{\partial y(x,t)}{\partial t} + \rho A\frac{\partial^2 y(x,t)}{\partial t^2} = f(x,t) \tag{41}$$

$$\sin\left(\frac{n\pi x}{l}\right)\left[\begin{array}{c}EI\left(\frac{n\pi}{l}\right)^4 T(t) + S\left(\frac{n\pi}{l}\right)^2 T(t)\\ + \beta I\left(\frac{n\pi}{l}\right)^4 T(t) + C\dot{T}(t) + \rho A\ddot{T}(t)\end{array}\right] = F\sin\omega_{dr}t \tag{42}$$

$$\ddot{T}(t) + \left[\frac{\beta I}{\rho A}\left(\frac{n\pi}{l}\right)^4 + \frac{C}{\rho A}\right]\dot{T}(t) + \left[\frac{S}{\rho A}\left(\frac{n\pi}{l}\right)^2 + \frac{EI}{\rho A}\left(\frac{n\pi}{l}\right)^4\right]T(t) = F\sin\omega_{dr}t \tag{43}$$

Expressing the model as a multi-degree system yields Eq. (44):

$$T(t) = Ae^{-\zeta\omega_n t}\sin\left(\omega_d t + \phi\right) + X\cos\left(\omega t - \theta\right) \tag{44}$$

The natural frequency of the power conductor under forced vibration is expressed in Eqs. (45) and (46) as:

$$\omega_n^2 = \frac{S}{\rho A}\left(\frac{n\pi}{l}\right)^2 + \frac{EI}{\rho A}\left(\frac{n\pi}{l}\right)^4 \tag{45}$$

$$2\xi\omega_n^2 = \left[\frac{\beta I}{\rho A}\left(\frac{n\pi}{l}\right)^4 + \frac{C}{\rho A}\right] \tag{46}$$

The solution to the equation of motion under forced vibration is expressed in Eq. (47) as:

$$y(x,t) = \sin\frac{n\pi x}{l}\left[Ae^{-\zeta\omega t}\sin\left(\sin\omega t + \phi\right) + X\cos\left(\omega t - \theta\right)\right] \tag{47}$$

6. Conductor self-damping and dampers

The influence of external and internal damping mechanisms was considered in the conductor vibration model. The factors considered included the following [1, 32]:

- The power conductor inter-strand motion and fluid damping. This is proportional to the conductor velocity and represented as viscous damping in the model.

- The rate of strain in the power conductor. This proportional to the internal damping of the power conductor.

The high voltage transmission line damped model is expressed in Eq. (48) as:

$$EI\frac{\partial^4 y(x,t)}{\partial x^4} - S\frac{\partial^2 y(x,t)}{\partial x^2} + \beta I\frac{\partial^5 y(x,t)}{\partial x^4 \partial t} + C\frac{\partial y(x,t)}{\partial t} + \rho A\frac{\partial^2 y(x,t)}{\partial t^2} = f(x,t) \qquad (48)$$

Where C and β represent damping constants. In the presence of axial load, viscous air damping, strain rate damping or Kelvin-Voigt damping, high voltage transmission line integrity can be managed.

There are various types of dampers that can be used to reduce vibration. The dampers are excited by the vibration of the power conductor and the vibration of their masses connected by the massager cable help to damp out energy. Stockbridge dampers are commonly installed on high voltage transmission lines to reduce aeolian vibrations. Stockbridge dampers can be symmetrical or asymmetrical in their design. An example of dampers installed on high voltage transmission lines is shown in **Figure 1**. The design of Stockbridge dampers follows the principle of cantilever beams with mass at the free ends. The contribution of dampers to power conductor vibration mitigation is to lower the severity of the vibration to a level that might prevent failure to the line.

Figure 1. Asymmetrical damper.

7. FEM MATLAB model setup, formulation and implementation

In order to implement the conductor model in MATLAB environment, finite element analysis formulation was done as function of the physical state of power transmission line conductor. The models developed using finite element analysis can then be implemented in MATLAB. The FEM model enables the analysis of the dynamic behavior and response of power line conductor to the dynamic forces of wind [33]. Consider a power transmission line subjected to dynamic aeolian vibration as an assembly of thin strands having distributed mass and elasticity. The physical model can be represented used partial differential equations. Each strand in the transmission line experiences axial, bending and torsional loads from the wind [34, 35]. The accurate representation of each factor is critical in the determination of the dynamic behavior of power transmission lines [36, 37]. Euler-Bernoulli curved beam theory was used to formulate the finite element model of power transmission lines.

Consider a power transmission line experiencing a vertical force, curvature and an axial force has an axial displacement modeled in Eq. (49) as [21]:

$$u(x,y) = u_0(x) + \frac{v}{R} - y\left(\theta_x + \frac{u(x)}{R}\right) \tag{49}$$

Where θ_x represents the rotation of the power line due to flexural effect, R represents the radius of rotation, y represents the distance from the axis of rotation to the centroidal axis of the conductor or transverse displacement, v represents tangential displacement and $u(x)$ represents the axial displacement of the power lines. The shape function for power transmission lines having rotation, bending and axial motion components is modeled using discretization techniques and represented in Eqs. (50)–(52) as:

$$u(s) = b_0 + b_1 s \tag{50}$$

$$v(s) = c_0 + c_1 S + c_2 S^2 + c_3 S^3 \tag{51}$$

$$\theta = \frac{dv(S)}{ds} = c_1 + 2c_2 S + 3c_3 S^2 \tag{52}$$

The solution to the discretization of the models yields Eq. (53) to Eq. (59):

$$\begin{bmatrix} u \\ v \\ \theta \end{bmatrix} = \begin{bmatrix} N_1 & 0 & N_2 & 0 & 0 & 0 \\ 0 & N_3 & 0 & N_4 & N_5 & N_6 \\ 0 & 0 & N_4 & 0 & N_5 & N_6 \end{bmatrix} \begin{bmatrix} u_1 \\ v_1 \\ \theta_1 \\ u_2 \\ v_2 \\ \theta_2 \end{bmatrix} \tag{53}$$

Where

$$N_1 = \frac{1}{2}(1 - \zeta) \tag{54}$$

$$N_2 = \frac{1}{2}(1 + \zeta) \tag{55}$$

$$N_3 = \frac{1}{4}(2 - 3\zeta + \zeta^3) \tag{56}$$

$$N_4 = \frac{1}{4}(1 - \zeta - \zeta^2 - \zeta^3) \tag{57}$$

$$N_5 = \frac{1}{4}(2 + 3\zeta - \zeta^3) \tag{58}$$

$$N_6 = \frac{1}{4}(1 - \zeta + \zeta^2 + \zeta^3) \tag{59}$$

The power line matrix model contains the strand stiffness K, mass matrix M and the load vector F. They are expressed in Eq. (60) as:

$$[K] = \frac{1}{2}\int N_a{}^T(EA)N_a \ \delta\zeta + \frac{1}{2}\int N_B{}^T(EI)N_B \ \delta\zeta + \frac{1}{2}\int N_B{}^T(T)N_{B,T} \ \delta\zeta \tag{60}$$

Where A represents the cross-sectional area of the power line, E represents the young modulus of the power line material, I represents polar moment of area, T represents the kinetic energy of the system. The matrix is modeled in Eq. (61).

$$[M] = \frac{1}{2}\int \dot{u}^T \rho A \dot{u} + \frac{1}{2}\int \dot{g}^T \rho A \dot{v} \tag{61}$$

Where ρ represents the density of the power line material and Eq. (62) indicates external excitation.

$$\delta W = \frac{1}{2}\int F \ \delta u \tag{62}$$

The power line conductor model is constructed using Euler-Bernoulli theories and summarized in Eq. (63) as:

$$\begin{bmatrix} M_{11} & M_{12} \\ M_{21} & M_{22} \end{bmatrix} \begin{pmatrix} \ddot{u} \\ \ddot{v} \end{pmatrix} + \begin{bmatrix} K_{11} & K_{12} \\ K_{21} & K_{22} \end{bmatrix} \begin{pmatrix} u \\ v \end{pmatrix} = \begin{pmatrix} F_1 \\ F_2 \end{pmatrix} \tag{63}$$

The finite element analysis follows a step by step numerical computation in the MATLAB environment as documented in [38, 39]. The dynamic response analysis assumes continuous displacement, velocity and acceleration [40, 41]. The numerical integration technique utilized was based on Newmark integration method. The compact form of the high voltage transmission line model is expressed in Eqs. (64)–(69) as [18]:

$$[M]\{\ddot{y}\} + [C]\{\dot{y}\} + [K]\{y\} = [F] \tag{64}$$

$$\left[\hat{K}\right]_{s+1} = [K]_{s+1} + a_3[m]_{s+1} \tag{65}$$

$$\left[\hat{F}\right]_{s,s+1} = \{F\}_{s+1} + [m]_{s+1}\left(a\{y\}_s + a\{\dot{y}\}_s + a\{\ddot{y}\}_s\right) \tag{66}$$

$$a_3 = \frac{2}{\gamma(\Delta t)^2} \tag{67}$$

$$a_4 = \frac{2}{\gamma \Delta t} \tag{68}$$

$$a_5 = \frac{1}{\gamma} - 1 \tag{69}$$

The initial conditions are expressed in Eq. (70) as:

$$[\ddot{y}]_0 = [M]^{-1}[F]_0 - [K]^{-1}[y]_0 \tag{70}$$

The acceleration vector is expressed in Eqs. (71)–(74) as:

$$[\ddot{y}]_{s+1} = a_3 \left(\{y\}_{s+1} - \{y\}_s \right) - a_4 \{\dot{y}\} - a_5 \{\ddot{y}\}_s \tag{71}$$

$$[\ddot{y}]_{s+1} = \{\ddot{y}\}_s + a_2 \{\ddot{y}\}_s = a_1 \{\ddot{y}\}_{s+1} \tag{72}$$

$$a_1 = \alpha \Delta t \tag{73}$$

$$a_2 = (1 - \alpha)\Delta t \tag{74}$$

8. FEM MATLAB model implementation strategy

In order to test the validity of the models discussed earlier using MATLAB, an aluminum power conductor with a steel core having a total diameter of 35.56 mm and having an ultimate tensile strength of 51.51kN was used in setting up the MATLAB simulation. Further physical properties of the power cable are shown in **Table 1**. The power conductor had a minimum bending stiffness EI_{min} of 8.66 Nm^2 and maximum bending stiffness EI_{max} of 433 Nm^2. The wholistic finite element models where implemented in MATLAB using strategy expressed in **Figure 2**.

Strand layer	Strand material	Diameter (mm)	No. of strands	Pitch per length (cm)	Lay direction
Layer 0	Steel	2.25	1		
Layer 1	Aluminum	3.38	6	16.1	Left hand lay
Layer 2	Aluminum	3.38	12	22.2	Right hand lay

Table 1. Power transmission conductor physical properties.

The inputs in the MATLAB algorithm were bending and axial loads, the cross-sectional area of the power conductor, strand radius and strand material type. The type of analysis which can be either static or dynamic was also specified as part initial and boundary conditions. Also included in the algorithm was to specify if the computation focuses on local vibration of the power conductor or the global vibration model.

9. Experimental investigation of conductor vibration

MATLAB code was written for the FEM and this was used to model the dynamic analysis of the problem of conductor vibration. To validate the FEM model an experimental study was conducted at the Vibration and Research Testing Centre (VRTC) situated at the University of KwaZulu-Natal which comprises of apparatus similar to that shown in **Figure 3**. The sweep tests (resonance search) were carried out and the test results were used to obtain natural frequencies and the modes of vibration for a Pelican conductor. The frequency range for the

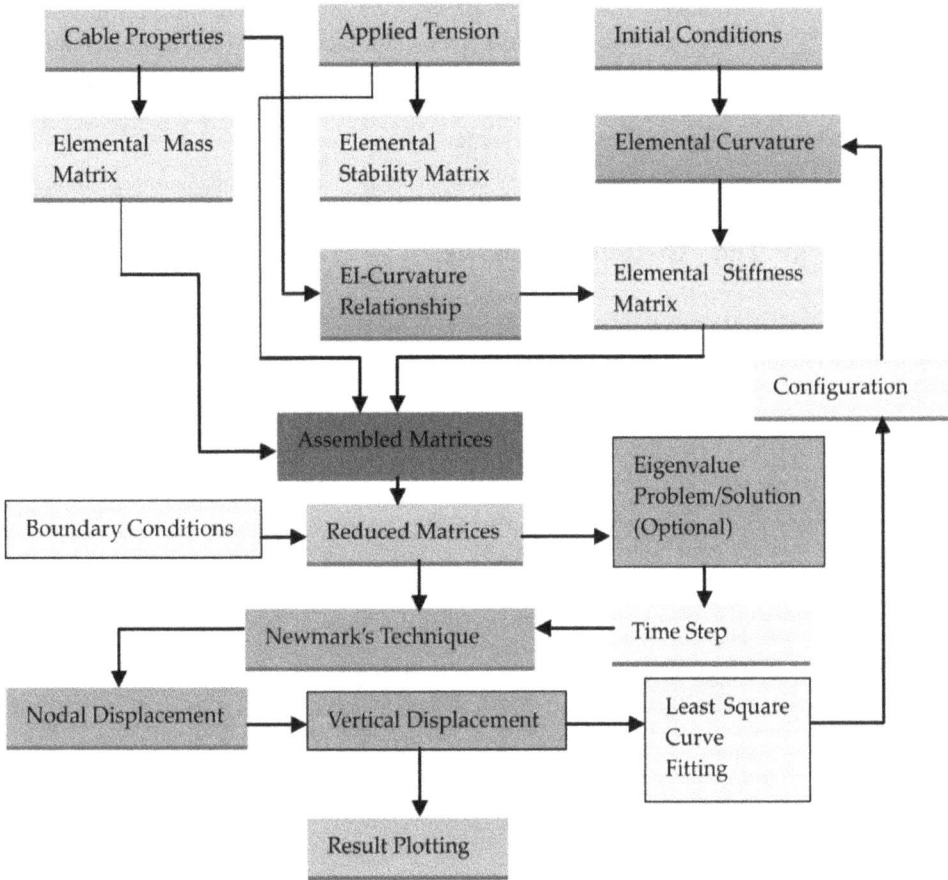

Figure 2. FEM MATLAB implementation strategy.

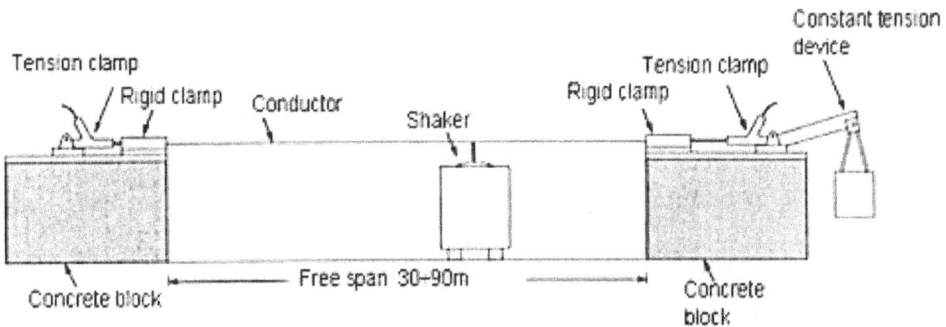

Figure 3. Experimental test set-up [7] .

Pelican conductor was between 5 and 50 Hz and testing was done for three axial tensions of 20, 25, 30 and 35% of its ultimate tensile strength (UTS). The experimental results obtained were used to validate the developed FEM model. The comparison between results from the experimental data, FEM and the theoretical model for the three different axial tensions for high voltage conductors are reported in the next section.

10. Simulation and experimental results

The results from the MATLAB simulations were compared with results from the finite element models (FEM) and experimental recordings. These are shown in **Figures 4–7**. The results were compared in terms of the natural frequency of vibration or vertical displacement of the power conductor.

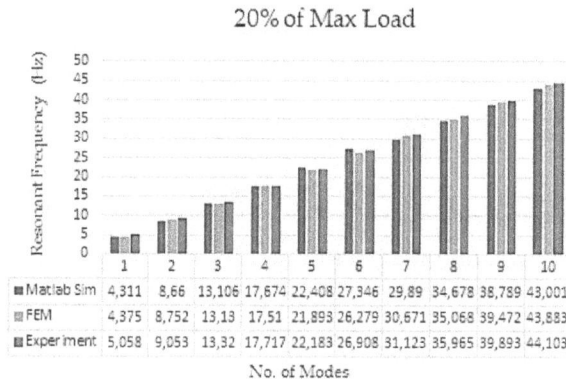

20% of Max Load

	1	2	3	4	5	6	7	8	9	10
■ Matlab Sim	4,311	8,66	13,106	17,674	22,408	27,346	29,89	34,678	38,789	43,001
▥ FEM	4,375	8,752	13,13	17,51	21,893	26,279	30,671	35,068	39,472	43,883
▦ Experiment	5,058	9,053	13,32	17,717	22,183	26,908	31,123	35,965	39,893	44,103

No. of Modes

Figure 4. Frequency of vibration at 20% UTS.

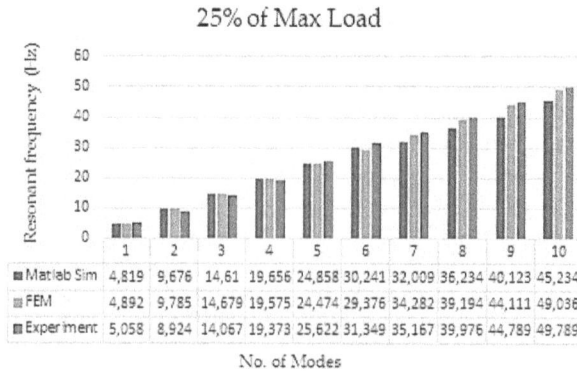

25% of Max Load

	1	2	3	4	5	6	7	8	9	10
■ Matlab Sim	4,819	9,676	14,61	19,656	24,858	30,241	32,009	36,234	40,123	45,234
▥ FEM	4,892	9,785	14,679	19,575	24,474	29,376	34,282	39,194	44,111	49,036
▦ Experiment	5,058	8,924	14,067	19,373	25,622	31,349	35,167	39,976	44,789	49,789

No. of Modes

Figure 5. Frequency of vibration at 25% UTS.

30% of Max Load

	1	2	3	4	5	6	7	8	9	10
Matlab Sim	5,277	10,59	15,973	21,461	27,087	32,882	37,002	42,123	47,689	53,211
FEM		5,3591	10,718	16,079	21,442	26,8	32,175	37,548	42,924	48,307	53,696
Experiment	5,692	10,014	14,734	19,934	25,11	30,283	34	38,987	43,895	48,789

No. of Modes

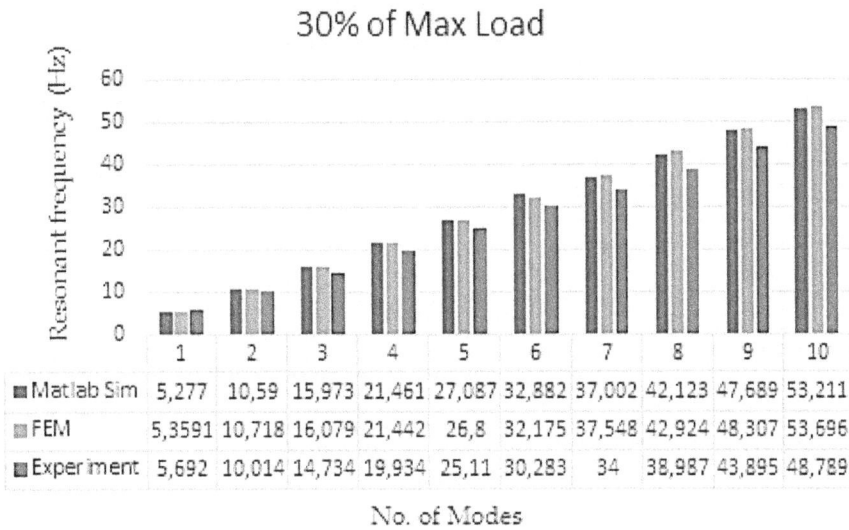

Figure 6. Frequency of vibration at 30% of UTS.

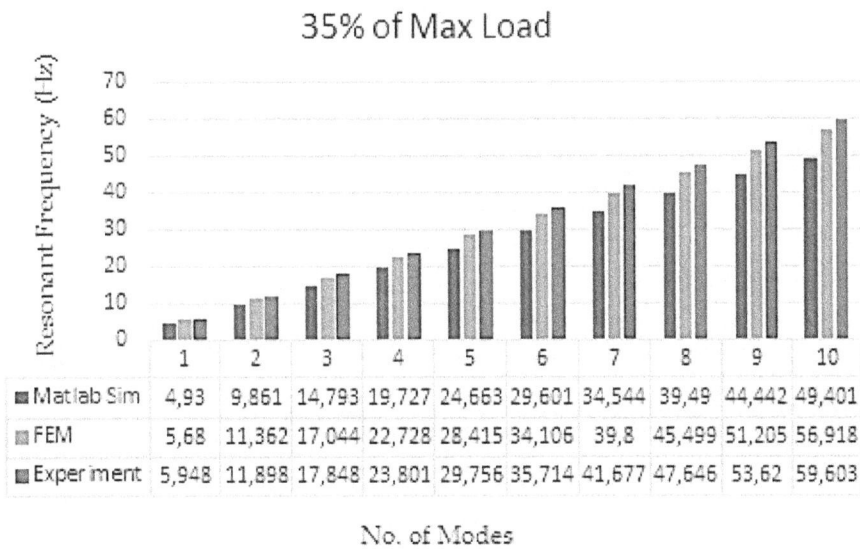

35% of Max Load

	1	2	3	4	5	6	7	8	9	10
Matlab Sim	4,93	9,861	14,793	19,727	24,663	29,601	34,544	39,49	44,442	49,401
FEM		5,68	11,362	17,044	22,728	28,415	34,106	39,8	45,499	51,205	56,918
Experiment	5,948	11,898	17,848	23,801	29,756	35,714	41,677	47,646	53,62	59,603

No. of Modes

Figure 7. Frequency of vibration at 35% of UTS.

11. Conclusion

The results showed that the implementation of the derived models in MATLAB provided a reliable strategy in the determination of the wind-induced dynamic properties of high voltage transmission lines. The results from MATLAB simulation, finite element method and experimental recordings were similar in values and showed similar trend. MATLAB as an environment can be used as a reliable simulation tool to implement and analyze high voltage conductor dynamics. The parameters obtained from the results, to some degree of accuracy can be used to predict the response of conductors due to aeolian vibration caused by wind loading.

Author details

Chiemela Onunka[1]* and Evans Eshiemogie Ojo[2]

*Address all correspondence to: onunka@mut.ac.za

1 Mangosuthu University of Technology, Durban, South Africa

2 Durban University of Technology, Durban, South Africa

References

[1] Ojo EE. Dynamic characteristics of bare conductors [thesis]. Durban: University of KwaZulu-Natal; 2011

[2] Guerard S, Godard B, Lilien J-L. Aeolian vibrations on power-line conductors, evaluations of actual self-damping. IEEE Transactions on Power Delivery. 2011;26(4):2118-2122

[3] Godard B, Guerard S, Lilien J-L. Original real-time observations of aeolian vibrations on power-line conductors. IEEE Transactions on Power Delivery. 2011;26(4):2111-2117

[4] Zhao L, Huang X. Integrated condition monitoring system of transmission lines based on fiber bragg grating sensor. In: Proceedings of the 2016 International Conference on Condition Monitoring and Diagnosis. 25-28 September, 2016; China, Xi'an: IEEE; 2016. p. 667-670

[5] Lalonde S, Guilbault R, Langlois S. Numerical analysis of ACSR conductor-clamp systems undergoing wind-induced cyclic loads. IEEE Transactions on Power Delivery. 2017; 99:1-9

[6] Lu ML, Chan JK. Rational design equations for the aeolian vibration of overhead power lines. In: Proceedings of 2015 IEEE Power & Energy General Meeting. 26–30 July, 2015; USA, Denver: IEEE; 2015. p. 1-5

[7] Langlois S, Legeron FLF. Time history modelling of vibrations on overhead conductors with variable bending stiffness. IEEE Transactions on Power Delivery. 2014;29(2):607–614

[8] Langlois S, Legeron F. Prediction of aeolian vibration on transmission-line conductors using a nonlinear time history model – Part I: Damper model. IEEE Transaction on Power Delivery. 2014;29(2):1168-1175

[9] Langlois S, Legeron F. Prediction of aeolian vibration on transmission-line conductors using a nonlinear time history model – Part II: Conductor and damper model. IEEE Transaction on Power Delivery. 2014;**29**(3):1176-1183

[10] Claren R, Diana G. Mathematical analysis of transmission line vibration. IEEE Transactions on Power Apparatus And Systems. 1969;**12**:1741-1771

[11] Hathout I, Callery-Broomfield K, Tang TT-T. Fuzzy probabilistic expert system for overhead conductor assessment and replacement. In: Proceedings of the 2015 IEEE Power & Energy Society General Meeting. 26–30 July, 2015; USA, Denver: IEEE; 2015. p.20-25

[12] Levesque F, Goudreau S, Langlois S, Legeron F. Experimental study of dynamic bending stiffness of ACSR overhead conductors. IEEE Transactions on Power Delivery. 2015;**30**(5): 2252-2259

[13] Alminhana F, Mason M, Albermani F. A compact nonlinear dynamic analysis technique for transmission line cascades. Engineering Structures. 2018;**158**:164-174

[14] El Damatty A, Elawady A. Critical load cases for lattice transmission line structures subjected to downbursts: Economic implications for design of transmission lines. Engineering Structures. 2018;**159**:213-226

[15] Barbieri N, Calado MKT, Mannala MJ, de Lima KY, Barbieri GSV. Dynamical analysis of various transmission line cables. Procedia Engineering. 2017;**199**:516-521

[16] Yin X, Wu W, Li H, Zhong K. Vibration transmission within beam-stiffened plate structures using dynamic stiffness method. Procedia Engineering. 2017;**199**:411-416

[17] Xie Q, Cai Y, Xue S. Wind-induced vibration of UHV transmission tower line system: Wind tunnel test on aero-elastic model. Journal of Wind Engineering & Industrial Aerodynamics. 2017;**171**:219-229

[18] Diana G, Falco M. On the forces transmitted to a vibrating cylinder by a blowing fluid. Meccanica. 1971;**6**:9-22

[19] Cigrè Study Committee 22-Working Group 01. Report on aeolian vibration. Electra. 1989;**1**(124):101

[20] Rawlins C.B. Model of power imparted to a vibrating conductor by turbulent wind [Report No. 93-83-3]. Spartanburg, South Carolina: Alcoa Conductor Products Company; 1983

[21] Deng HZ, Xu HJ, Duan CY, Jin XH, Wang ZH. Experimental and numerical study on the responses of a transmission tower to skew incident winds. Journal of Wind Engineering & Industrial Aerodynamics. 2016;**157**:171-188

[22] Ghabraei S, Moradi H, Vossoughi G. Finite time-Lyapunov based approach for robust adaptive control of wind-induced oscillations in power transmission lines. Journal of Sound and Vibration. 2016;**371**:19-34

[23] EPRI. Transmission line reference book: Wind-induced conductor motion. Electrical Power Research Institute. Palo Alto, USA: EPRI; 1979

[24] EPRI. Transmission line reference book: Wind-induced conductor motion final report. Palo Alto, USA: EPRI; 2016

[25] Vecchiarelli J, Currie I, Havard D. Computational analysis of aeolian conductor vibration with a stockbridge-type damper. Journal of Fluids and Structures. 2000;**14**:489-509

[26] Hong K-J, Der Kiureghian A, Sackman JL. Mint: Bending behavior of helically wrapped cables. Journal of Engineering Mechanics. 2005;**131**(5):500

[27] Gizaw M, Davidson IE, Loubser R, Bright G, Stephen R. Analyses of the vibration level of an OPGW at catenary value of 2100 m with multi-response Stockbridge dampers. In: Proceedings of the 2016 IEEE PES Power Africa Conference. June 28–July 2, 2016, Zambia, Livingstone: IEEE; 2016. p. 107-111

[28] Xiao S, Wang H, Ling L. Research on a novel maintenance robot for power transmission lines. In: Proceedings of 4th International Conference on Applied Robotics for the Power Industry. 11–13 October, 2016; China, Jinan: IEEE; 2016. p. 1-6

[29] Mathworks. MATLAB. The Mathworks Inc. 2017. [Online]. Available from: https://www.mathworks.com/products/matlab.html. [Accessed: Nov 11, 2017]

[30] Xie T, Peng Z, Zhou Z. Study on optimization of anti-corona properties of 330-kv dampers. IEEE Transactions on Power Delivery. 2015;**30**(4):1827-1832

[31] Lalonde S, Guilbault R, Langlois S. Modelling multilayered wire strands, a strategy based on 3D finte element beam-to-beam contacts – Part II: Application to wind-induced vibration and fatigue analysis of overhead conductors. International Journal of Mechanical Sciences. 2017;**126**:297-307

[32] Hardy C. Analysis of self-damping characteristics of stranded cables in transverse vibration. In: Proceedings of CSME Mechanical Engineering Forum. 3–9 June, 1990; Canada, Toronto: CSME; 1990. p. 117-122

[33] Ojo EE, Ijumba NM. Mint: Numerical method for evaluating the dynamic behaviour of power line conductors: A global approach for pure bending. International Journal of Engineering Research & Technology (IJERT). 2016;**5**(3):584-589. ISSN: 2278-0181

[34] Lanteigne J. Theoretical estimation of the response of helically armored cables to tension, torsion, and bending. Journal of Applied Mechanics. 1985;**52**:423-432

[35] IEC 62219. Overhead electrical conductors-formed wire, concentric lay and stranded conductors. IEC. 2002;**1**:1-41

[36] Jiang W, Wang T, Jones W. Forced vibration of coupled extensional-torsional systems. Journal of Engineering Mechanics. 1991;**117**:1171-1190

[37] Ojo EE. Finite element formulation and analysis of the composite structure of overhead transmission lines conductors. In: Proceedings of the 10th South African Conference on Computational and Applied Mechanics. 3–5 October, 2016; South Africa, Potchefstroom: SACAM; 2016. p. 382-393

[38] Ojo EE, Shindin S. Mint: Finite element analysis of the dynamic behaviour of transmission line conductors using MATLAB. Journal of Mechanics Engineering and Automation. 2014;**4**:142-148

[39] Kwon YW, Bang H. Finite Element Method Using Matlab. 2nd ed. Boca Raton: CRC Press; 2000

[40] Wilson EL, Clough RW. Dynamic response by step-by-step matrix analysis. In: Proceedings of the Symposium on the Use of Computers in Civil Engineering. 1–5 October, 1962; Portugal, Lisbon: CICE; 1962. p. 1-14

[41] Newmark MN. A method of computation for structural dynamics. Journal of Engineering Mechanics Division and ASCE Proceedings. 1959;**85**:EM3

www.ingramcontent.com/pod-product-compliance
Lightning Source LLC
Chambersburg PA
CBHW081245190326
41458CB00016B/5926

* 9 7 8 1 7 8 9 2 3 7 0 6 1 *